T0135975

SCIENCE-REPORT

aus dem Faserinstitut Bremen | Band 4

Mohammad Mokbul Hossain

Plasma Technology for

Deposition and Surface Modification

Herausgeber:
Prof. Dr.-Ing. Axel S. Herrmann

Science-Report aus dem Faserinstitut Bremen

Hrsg.: Prof. Dr.-Ing. Axel S. Herrmann

ISSN 1611-3861

Bibliografische Information der Deutschen Nationalbibliothek

Die Deutsche Nationalbibliothek verzeichnet diese Publikation in der Deutschen Nationalbibliografie; detaillierte bibliografische Daten sind im Internet über http://dnb.d-nb.de abrufbar.

ISBN 978-3-8325-2074-8

Logos Verlag Berlin GmbH
Comeniushof, Gubener Str. 47,
10243 Berlin
Tel.: +49 030 42 85 10 90
Fax: +49 030 42 85 10 92
INTERNET: http://www.logos-verlag.de

Fibers
Advanced

In Zusammenarbeit mit:
Empa, Swiss Materials Science & Technology

Betreuer:
Prof. Dr.-Ing. Axel S. Herrmann
Dr. rer. nat. Dirk Hegemann

Materials Science & Technology

To my mother and father
with love

ACKNOWLEDGEMENTS

I would like to thank all the individuals and institutions for their kind cooperation in providing me with valuable information, feedback and support in the creation of this dissertation. Without their help and support this dissertation would have not crystallized to this wonderful stage.

I gratefully acknowledge Professor Axel S. Herrmann from the Fiber Institute Bremen at the University of Bremen, Germany for giving me the opportunity to work in his group and for supervising me throughout this research.

Most of all, I would like to thank Empa, Swiss Materials Science & Technology for welcoming me as a PhD student in the Laboratory for Advanced Fibers and allowing me to enjoy its wonderful research opportunities and well-equipped lab facilities.

I express my heartfelt gratitude to my guide Dr. Dirk Hegemann for his inspiring guidance and supervision. I sincerely thank him for giving me all the necessary help to undertake this work.

I am very thankful to Professor Jörg Müssig for his support and collaborative work with the Fiber Institute Bremen, in particular for work associated with fiber-reinforced composites.

A special thanks goes to Dr. Manfred Heuberger and Dr. Rudolf Hufenus for their valuable suggestions and beneficial discussions.

A big thanks goes to Mr. Martin Amberg for his technical help and support. I would like to further thank former colleagues Mr. Peter Furrer for the dyeing tests and Mr. Urs Schutz for his help.

I take this opportunity to thank both researchers Dr. Giuseppino Fortunato and Dr. Jörn Lübben for their valuable contributions to my thesis, in particular for the XPS and AFM analysis, respectively.
A very special thanks goes to Dr. Axel Ritter for collaborative work concerning plasma and wet chemistry. I would like to further thank Dr. Ratnesh Thapliyal, Dmitry Nazarov, Oshiorenoya Agabi, Thomas Ruddy and Dr. Dawn J. Balazs for their help and encouragement.

I am indebted to my thesis committee, Professor Georg Grathwohl from the Institute of Ceramic Materials and Components, Professor Reinhold Kienzler from the Department of Technical Mechanics - Structural Mechanics and Mr. Christoph Hoffmeister from the Fiber Institute Bremen, who kindly agreed to be co-examiners in the jury.

I would like to express my sincere gratitude to all other colleagues in the plasma group at Empa for all the interesting and fruitful meetings, discussions, valuable guidance, and for the wonderful years we spent together.

I greatly appreciated our secretary Ms. Brigitte Knöpfel van der Niepoort and secretary Ms. Eva Gleitze from the Fiber Institute Bremen for their kindness and assistance.

I would like to acknowledge the financial support from Empa, as well as the Fiber Institute Bremen. I would also like to acknowledge the Commission for Technology and Innovation (CTI), Bern – Switzerland for funding.

Last but not least, I would like to thank my parents for their love and encouragement.

ABSTRACT

Plasma processing is a high-technology discipline in tailoring surface properties and in obtaining functional polymers of advanced materials without changing the material's bulk. Comparing with solid polymeric materials, special care should be taken for surface activation of textiles due to their complex geometries. It was found that modification is strongly influenced by both plasma parameters and fabric structure. As compared to air, CO_2, and water vapor, Ar/O_2 and He/O_2 mixtures were found to be very effective for surface hydrophilization of polyester (PES) textiles due to the long-lasting free radical lifetimes. Moreover, the degree of plasma species penetration and the permanency of the treatment are closely linked to weave construction. In particular, the hydrophilicity of looser structured fabrics is enhanced remarkably as compared to tightly woven fabrics. The wettability of plasma-treated PES fabrics is improved significantly due to the formation of polar groups on the surface. The modified surfaces were not stable for a long time due to restructuring of the polar functional groups. Therefore, plasma coatings containing functional groups are required in order to obtain a permanent surface modification.
Besides plasma parameters the gas ratio, in particular when non-polymerizable gases are mixed with polymerizable gases, plays a very important role in determining mass deposition rates. The following macroscopic approach was established in order to determine the deposition rates of different gas mixtures (O_2/HMDSO, N_2 as well as NH_3/C_xH_y, and inert gas/C_xH_y). By modifying the well-known reaction parameter *W/F* (specific energy) and by introduction of a modified flow $F = F_m + a\ F_c$ (sum of monomer flow F_m and carrier reactive gas flow F_c with a flow factor *a*) the influence of the reactive gas on plasma polymerization can be identified. Additionally, it is promising to optimize and scale-up plasma polymerization processes.
Permanent nanoporous coatings were deposited in order to obtain functional surfaces which contain accessible functionalities within the entire coating volume. This novel approach is essentially based on a fine control of simultaneous deposition and etching processes during plasma co-polymerization of ammonia with hydrocarbons (C_2H_2 and C_2H_4) at room temperature. A nanoporous structure with a large specific surface area was achieved that contained functional groups inside the coating volume, which were accessible to e.g. dye molecules, thus facilitating substrate independent dyeing. The coloration of plasma polymers provides information about the accessible amine functionalities, coating purity and uniformity. Therefore, the coloration can be used as a chemical tracer for the nanoporous coatings. Likewise, the coatings can be exploited to attract biomolecules important to cell-adhesion. Varying deposition conditions were applied in order to tune both the nitrogen functionalities and porosity in the nanocomposites.
A permanent hydrophilic modification of material surfaces was obtained by introducing nitrogen polar functionalities, depending on the NH_3 to hydrocarbon ratio, which is mostly due to a replacement of carbon in a-C:H:N films. This novel combination of polar groups with a suitable texturing realized within crosslinked a-C:H:N coatings proved to be an efficient method providing a long-term mechanical stability of superhydrophilic coatings. It was evident that the coating quality could be improved significantly using a C_2H_4/NH_3 plasma due to reduced unsaturated bonds. A high dyeing fastness indicated a strong dye-molecule bonding indicating

permanency of the amines. Furthermore, since the coating thickness is in the nanometer range (<100 nm), the materials, architectural porosity, touch, and comfort etc. are not affected.
Plasma coated material surfaces, e.g. fibers, contain huge numbers of functional groups such as amines, carboxyl, hydroxyl etc. which can chemically interact with matrix materials and hence, yield strong covalent bond between fiber and matrix. The coatings show a large surface area which enhances the contact area and surface texturing and additionally promotes mechanical interlocking. As a result, interfacial adhesion and strength between fiber and matrix were greatly improved. The results showed that superior interfacial adhesion was obtained for nanoporous coated fibers as compared to untreated fibers. Thus, biodegradable natural fiber-reinforced "green composites" can be obtained, which satisfy sustainability criteria.
In addition, since the plasma treatment is largely independent of the substrate material, this suggests the possibility of using a universal coating process instead of optimizing surface modification processes and plasma parameters for each different substrate material. The technique represents a straightforward industrial process to develop tailor-made multifunctional nanoporous composites.
Thus, the novel, developed nanoporous coatings incorporating accessible functional groups represent a platform for diverse multifunctional applications in the surface enhancement of advanced materials.

VERSION DEUTSCH

Die Plasmatechnologie zur Behandlung von Materialoberflächen stellt eine Hightech-Methode dar, bei der Oberflächeneigenschaften angepasst und funktionelle Plasmapolymere abgeschieden werden können, ohne dabei die Substrateigenschaften zu verändern. Im Gegensatz zu polymeren Formkörpern sollte die Oberflächenaktivierung von Textilien aufgrund der komplexen Geometrie mit besonderer Sorgfalt durchgeführt werden. Dabei stellt sich heraus, dass die Funktionalisierung der Textilien stark durch die Plasmaparameter sowie durch die Gewebestruktur beeinflusst wird. Im Vergleich der Reaktivgase Luft, CO_2 und Wasserdampf erwiesen sich Ar/O_2- und He/O_2-Mischungen bei der Oberflächenbenetzung von Polyester (PES)-Textilien, dank der grossen Anzahl langlebiger Radikalen, als sehr effektiv. Weiterhin ist die Eindringtiefe der Plasmabehandlung und die Permanenz der Behandlung eng mit der Gewebestruktur verknüpft. Besonders die Benetzbarkeit von offenen strukturierten Geweben wird gegenüber der an geschlossenen Geweben erreichbaren beachtlich gesteigert. Durch Bildung polarer Gruppen auf der Oberfläche wird der Benetzungsgrad plasmabehandelter PES-Gewebe merklich verbessert. Aufgrund der Restrukturierung polarer, funktioneller Gruppen sind die modifizierten Oberflächen über einen längeren Zeitraum nicht stabil. Daher sind vernetzte Plasma-Beschichtungen mit zugänglichen, funktionellen Gruppen notwendig, um eine permanente Oberflächen-Änderung zu erzielen.
Neben den Plasmaparametern spielt das Gasverhältnis eine wichtige Rolle beim Bestimmen der Massenabscheidungsrate, besonders bei der Mischung von nicht-polymerisierbaren mit polymerisierbaren Gasen. Mit Hilfe der nachstehend beschriebenen makroskopischen Verfahrensweise wurden die Abscheidungsraten verschiedener Gase (O_2/HMDSO sowie Kohlenwasserstoffe mit verschiedenen Reaktivgasen) systematisch untersucht. Indem der gut bekannte Reaktions-Parameter Leistungseintrag pro Fluss W/F (spezifische Energie) geändert und ein modifizierter Fluss $F = F_m + a\ F_c$ (Summe des monomeren Flusses F_m und des Trägergas-Flusses F_c mit Wichtungsfaktor a) eingeführt wird, können schichtbildende Prozesse identifiziert werden. Zusätzlich wird es möglich, die Plasmapolymerisation zu optimieren und zu skalieren.
Nanoporöse Beschichtungen wurden auf Textilien aufgetragen, um funktionelle Oberflächen mit zugänglichen funktionellen Gruppen innerhalb des gesamten Beschichtungsvolumens zu erzielen. Diese innovative Methode stützt sich hauptsächlich auf die feine (genau) Kontrolle gleichzeitiger Abscheidungs-/Ätzprozesse während der Plasma-Copolymerisierung bei niedriger Temperatur von Ammoniak mit Kohlenwasserstoffen (C_2H_2 und C_2H_4). Eine nanoporöse Struktur mit einer grossen spezifischen Oberfläche wurde derart hergestellt. Diese Beschichtung enthält funktionelle Gruppen innerhalb der Plasmaschicht, die z.B. für Farbstoff-Moleküle zugänglich sind und somit das substratunabhängige Färben von Oberflächen ermöglicht. Die Farbintensität der gefärbten Plasma-Polymere gibt z.B. Auskunft über die zugänglichen Amin-Funktionalitäten, die Beschichtungsgüte und Gleichmässigkeit.
Eine dauerhafte hydrophile Änderung der Materialoberflächen wurde in Abhängigkeit von NH_3 zu C_xH_y Verhältnis durch die Einführung von Stickstoff-polaren Funktionalitäten erzielt, die hauptsächlich auf die Substitution von Kohlenstoff in a-

C:H:N-Schichten zurückzuführen ist. Diese innovative Kombination polarer Gruppen mit passender Texturierung innerhalb vernetzter a-C:H:N-Beschichtungen erwies sich als eine wirksame Methode, die superhydrophilen Beschichtungen eine dauerhafte Stabilität verleiht. Es war eindeutig, dass die Beschichtungsqualität mit Hilfe eines C_2H_4/NH_3-Plasmas dank der reduzierten Anzahl ungesättigter Bindungen beachtlich verbessert werden konnte. Hohe Farbechtheit weist auf eine starke Farbmolekülbindung hin, welche die Verfügbarkeit der Amin-Gruppen aufzeigt. Da sich die Schichtdicke jeweils innerhalb des Nanometer-Bereichs (<100 nm) befindet, werden Gewebekomfort, Griff, Luftdurchlässigkeiten usw. nicht beeinflusst.

Plasmabeschichtete Faseroberflächen enthalten hohe Mengen an funktionellen Gruppen, wie Amine, Carboxyl, Hydroxyl usw., die chemisch mit Matrix-Materialien zusammenwirken können. Dabei entsteht eine starke kovalente Bindung zwischen Faser und Matrix. Die Beschichtungen weisen hohe spezifische Oberflächen auf, die eine Erweiterung des Kontaktbereichs sichern, und die Oberflächentexturierung fördert zudem ein mechanisches Ineinandergreifen. Folglich konnten Grenzflächenhaftung und Festigkeit zwischen Faser und Matrix in grossem Maße verbessert werden. Die Resultate zeigen, dass gegenüber unbehandelten Fasern eine höhere Grenzflächenhaftung für nanoporös beschichtete Fasern erreicht wurde. Demzufolge können nachhaltige, bioabbaubare “grüne Naturfaser-Verbundstoffe” hergestellt werden, die gute Aussichten für ökologische und umweltfreundliche Lösungen bieten.

Innovativ entwickelte nanoporöse Beschichtungen, in denen zugängliche funktionelle Gruppen enthalten sind, können somit als Grundlage für verschiedene multifunktionelle Anwendungen in der Oberflächenverbesserung fortschrittlicher Materialien dienen.

LIST OF ABREVIATIONS AND SYMBOLS

a-C:H	amorphous hydrogenated carbon
a-C:H:N	nitrogenated amorphous hydrocarbon
AFM	Atomic Force Microscopy
BET	Brunauer, Emmett & Teller method
CA	contact angle
CASSING	crosslinking by activated species by inert gases
CH_4	methane
C_2H_2	acetylene
C_2H_4	ethylene
C_xH_y	hydrocarbon
CIE	Commission Internationale d'Eclairage (international body for colorimetry)
DC	direct current
*DE**	color difference
DLC	diamond-like
E_a	activation energy
ESCA	Electron Spectroscopy for Chemical Analysis
FC	fluorocarbon
F_c	carrier or reactive gas flow
F_m	monomer flow
g/m^2	gram per square meter
HMDSO	hexamethyldioxysiloxane
IR	infrared
K/S	color intensity
LPD	low pressure plasma
L:R	liquor to fabric ratio
MW	microwave
MW-ECR	Microwave Electron Cyclotron Resonance
NCM	non-contact mode
NaOH	sodium hydroxide
NH_3	ammonia
o.w.f	on the weight of fabric
Pa	pascal
PDMS	polydimethylsiloxane
PE	polyethylene
PET	poly(ethylene terephthalate)
PE-CVD	plasma-enhanced chemical vapor deposition
PEO	poly(ethylene oxide)
PES	polyester
PLA	poly(lactic acid)
PP	polypropylene
RF	radio frequency
R_m	mass deposition rate
R_{pp}	peak to peak roughness
R_{rms}	root mean square roughness
RSGP	rapid step-growth polymerization

R%	reflectance in percent
sccm	standard cubic centimeters per minute
SN	Swiss Norm
UV	ultraviolet
V/I	voltage current
VUV	vacuum ultraviolet
W/F	watt per flow (sccm) (energy input)
XPS	X-ray Photoelectron Spectroscopy
μCT	X-ray Microtomography

TABLE OF CONTENTS

1. Introduction

1.1. Scope of the Dissertation

Plasma, the fourth state of matter, where species such as neutral particles, electrons, positive and negative ions and photons, and molecules in ground and excited states coexist together, has emerged as an important technology in materials science since 1960s.[1] Plasma can be accelerated and steered by electric and magnetic fields, which allows it to be controlled and applied. Thus plasma, as a very reactive medium, now-a-days has been utilized as an important tool for materials surface processing, e.g. to modify the surface of a certain substrate, to deposit chemical materials on top of the substrate surface to impart desired properties, to remove substances which were previously deposited on the substrate.[2,3] The main advantage of these methods is that they almost exclusively modify the polymer substrate surface to a depth of approximately 500 Å, thus bulk material properties remains unaffected.[4]

When manufacturers scramble to find processes of surface tailoring that ensure successful advanced technology products for polymers, plastics, metals, glass, textiles with a variety of choices, plasma technology becomes a powerful tool in industrial processing over wet chemical methods. The main reason for the increasing interest is that industrially well-established surface-finishing processes suffer considerably from environmental and societal demands.[5] Moreover, plasma is able to process many different gaseous, liquid or solid precursors contrary to chemical processes/wet chemistry, since it is not limited to the reactivity of these substances at a given temperature or surface conditions.[6] It can also synergistically combine chemical reactivity with energetic bombardment of the surface. Plasma polymer does not contain regularly repeating units, the chains are branched and are randomly terminated with a high degree of crosslinking.[7,8] They adhere well to the solid surfaces.[9] Thus, plasma polymerization can be thought of as a “tool box” of technologies, which can provide surface solutions for the deposition of solid polymer films to a wide range of materials and application.[10,11]

However, the versatility is a result of its complexity, where phenomena as plasma physics, process design and control, plasma and surface chemistry, material science and electrical engineering are combined together, requiring an interdisciplinary study to plasmas.[12-17] The processing evolves/advances from the macroscopic scale to nanoscaled description in tailoring process conditions to meet objectives. That is the reason why applications precede the understanding from beginning of the adoption of plasma technology.

Plasma technology is increasingly attracting the textile world because it is widely perceived as offering huge potential in a wide spectrum of modifications.[18,19] This is mainly due to the fact that conventional methods used to impart different properties to fabrics do not lead to permanent effects, and will lose their functions after laundering or wearing. Due to the complex geometry of textiles, the application of plasma to textiles is even more complicated.[20,21] Surface activation of textiles, such as hydrophilic, hydrophobic etc., by plasma have been extensively investigated in numerous literatures,[22-25] but the modification does not provide durability for fabrics due to functional group reorganization on the modified surfaces.[26-28] Therefore, to obtain a permanent surface functionalization the deposition of plasma coatings is required.[29] Consequently, a large surface area with a high number of functional group containing coatings can be obtained, thus presenting better affinity for fabrics and leading to an increase in durability of the function. Since the functionalities in the coatings are accessible, the coatings can be used as foundation for various technological applications such as super-hydrophobicity, super-hydrophilicity, protein adsorption, UV protection, interfacial adhesion of fiber-reinforced composites etc. In addition, nanoscaled coatings on fabrics will not affect their breathability or hand feel.

1.2. Outline

In the scope of this dissertation, the work presented investigates the processes in order to obtain permanent nanoscaled coatings on textiles. For the surface activation, various non-polymer forming gaseous plasmas were extensively investigated and process parameters were optimized, thus promoting bonding between textiles and subsequent nanoporous plasma coatings. The evaluation of changes in surface chemistry and surface wettability were characterized using mainly the surface characterization techniques X-ray Photoelectron Spectroscopy (XPS) and contact angle (CA) measurements. Accessible functionalities in the nanoscaled coatings were examined by coloration and Atomic Force Microscopy (AFM). The wear and life time of the coatings were evaluated by washing test and abrasion tests. The chapters are organized as follows:

Chapter 2 describes plasma theory, and outlines the surface modification methods of plasma treatment, plasma polymerization and surface analysis methods.

Chapter 3 describes the optimization of plasma process parameters for the surface activation by non-polymer forming gases.

Chapter 4 explains the effects of plasma species penetration into different textile structures using argon with oxygen gaseous plasmas.

Chapter 5 describes the growth mechanism of plasma polymerization and the influence of reactive gases added during plasma polymerization.

Chapter 6 presents a systematic study of the deposition parameters for the nanoporous coatings using ammonia/acetylene gaseous plasmas.

Chapter 7 presents the deposition of nanoporous coatings using ammonia/ethylene gaseous plasmas.

Chapter 8 describes comparison studies between ammonia/acetylene and ammonia/ethylene plasmas and their coating quality characterized by coloration.

The last modification is presented in **Chapter 9** and describes permanent superhydrophilic thin films on textiles.

The final chapter, **Chapter 10**, provides potential and feasible application possibilities for the developed nanoporous thin coatings in textiles, followed by the concluding remarks.

1.3. References

[1] K. Samanta, M. Jassal, A. Agrawal, *Indian J. Fibre Text.* **2006**, *31*, 83.
[2] G. Bonizzoni, E. Vassallo, *Vacuum* **2002**, *64*, 327.
[3] M. Laroussi, C. Tendero, X. Lu, S. Alla, W. L. Hynes, *Plasma Process. Polym.* **2006**, *3*, 470.
[4] P. Heyse, R. Dams, S. Paulussen, K. Houthoofd, K. Janssen, P. A. Jacobs, B. F. Sels, *Plasma Process. Polym.* **2007**, *4*, 145.
[5] M. G. McCord, Y. J. Hwang, Y. Qiu, L. K. Hughes, M. A. Bourham, *J. Appl. Polym. Sci.* **2003**, *88*, 2038.
[6] M. J. Tsafack, J. Levalois-Grützmacher, *Surf. Coat. Technol.* **2007**, *201*, 5789.
[7] R. Balkova, J. Zemek, V. Cech, J. Vanek, R. Prokryl, *Surf. Coat. Technol.* **2003**, *174-175*, 1159.
[8] K. S. Siow, L. Britcher, S. Kumar, H. J. Griesser, *Plasma Process. Polym.* **2006**, *3*, 392.
[9] S. Gaur, G. Vergason, *Society of Vacuum Coaters*, 43rd Ann. Techn. Conf. Proceed., Denver, April 15-20, **2000**.
[10] G. S. Kakad, A. R. Rathod, B. Suman, *Man-made Text. India* **2006**, *4*, 85.
[11] B. Borer, A. Sonnenfeld, Ph. R. von Rohr, *Surf. Coat. Technol.* **2006**, *201*, 1757.
[12] K. Ostrikov, H. Yoon, A. E. Rider, S. V. Vladimirov, *Plasma Process. Polym.* **2007**, *4*, 27.
[13] S. Morita, S. Hattori, *Pure Appl. Chem.* **1985**, *57 (9)*, 1277.
[14] S. Zanini, C. Riccardi, M. Orlandi, P. Esena, M. Tontini, M. Dilani, V. Cassio, *Surf. Coat. Technol.* **2005**, *200*, 953.
[15] J. Benedikt, "Acetylene Chemistry in Remote Plasmas", PhD Thesis **2004**, Technische Universität Eindhoven, Netherland.
[16] F. F. Chen, *Phys. Plasmas* **1995**, *2 (6)*, 2164.
[17] M. R. Wertheimer, H. R. Thomas, M. J. Perri, J. E. Klemberg-Sapieha, L. Martinu, *Pure Appl. Chem.* **1996**, *68 (5)*, 1047.
[18] J. Verschuren, *Industry-Ready Innovative Research*, 1st Flanders Engineering PhD Symposium, Brussels, December 11, **2003**.
[19] W. J. Thorsen, *Text. Res. J.* **1974**, *6*, 422.
[20] M. Kabajev, I. Prosycevas, G. Kazakeviciute, V. Valiene, *Mat. Sci (Medziagotyra)* **2004**, *10 (2)*, 173.
[21] U. Vohrer, M. Müller, C. Oehr, *Surf. Coat. Technol.* **1998**, *98*, 1128.
[22] L. V. Sharnina, *Fibre Chem.* **2004**, *36 (6)*, 431.
[23] M. Lee, T. Wakida, M. S. Lee, P. K. Pak, J. Chen, *J. Appl. Polym. Sci.* **2001**, *80*, 1058.
[24] B. Kutlu, A. Cireli, "Plasma Technology for Textile Processing", Dokuz Eylul University, 35100, Borniva-IZMIR.
[25] R. Molina, P. Jovancic, D. Jocic, E. Bertran, P. Erra, *Surf. Interface Anal.* **2003**, *35*, 128.
[26] G. Placinta, F. Arefi-Khonsari, Gheorghiu, J. A. Ouroux, G. Popa, *J. Appl. Polym. Sci.* **1997**, *66*, 1367.
[27] M. M. Hossain, A. S. Herrmann, D. Hegemann, *Plasma Process. Polym.* **2006**, *3*, 299.
[28] M. M. Hossain, D. Hegemann, P. Chabrecek, A. S. Herrmann, *J. Appl. Polym. Sci.* **2006**, *102*, 1452.

[29] M. M. Hossain, A. S. Herrmann, D. Hegemann, *Plasma Process. Polym.* **2007**, *4*, 135.

2. Plasma Treatment

2.1. Plasma and Their Properties

The plasma state is generated when a gas is subjected to sufficient energy to break its molecular integrity and dissociate it into ions, electrons and other subatomic species. Langmuir was the first to use the term "plasma" in 1928 to describe the inner region of an electrical discharge.[1] Later the definition was broadened to define a state of matter, i.e. "ionized gas", in which a significant number of atoms or molecules are electrically charged or ionized, and it is referred to as the fourth state of matter.[2,3] An ionized gas consists mainly of positively charged (ionized) molecules or atoms (ions) and negatively charged electrons. The modern definition of plasma is given by Yasuda[4]: "plasma simply denotes a more or less ionized gas". A useful definition of the plasma state has been postulated by Chen[5] "a plasma is a quasi-neutral gas of charged particles which exhibits a collective behavior".

Plasma exists everywhere in nature as "natural plasma"; most of the material in the visible universe, as much as 99% according to some estimates, consists of matter in the plasma state. For example, the ionization is caused by high temperature at about 10^7 K (inside the sun). In the case of man-made gaseous plasma, a steady electric field is created between a metal electrode and the outer plasma reactor by applying an electric voltage to the electrode. Under these conditions, electrons jump from the electrode into the gas within the reactor. Simultaneously, a changing electrical field is created inside the reactor by the oscillating electric voltage on the electrode. Changing electrical fields produce changing magnetic fields. Free electrons initiate the process; exposure to an external energy source allows one to gain sufficient kinetic energy, so that a collision with another atom or molecule will result in the formation of ions and radicals. The reactive radical species are capable of chemical work where the ionized atom and molecular species are capable of physical work through sputtering by the emission of atoms/clusters. All the collisions with energetic particles readily move atoms and ions from their ground atomic level to an excited atomic level. The excited states relax back to their ground levels by emitting a burst of light energy known as a photon, thus, plasma is a visible glow. The wavelength of the emission is sufficient to break chemical bonds.

Since plasma is a good conductor, electric potentials play an important role.[6] The electric potential in most of the space between the electrodes is nearly constant. Considering the distance along the axis of a tube as x,

$$dV/dx = \text{const}, \quad d^2V/dx^2 = 0$$

$$d^2V/dx^2 = -4\pi\rho, \text{ thus } \rho = 0, \text{ i.e. } +ve = -ve$$

where, ρ the space charge density originates from the difference in density of the positive charge and the negative charge. Thus, positive charge density is equal to negative charge density. This results in the important concept of quasineutrality; but on the scale of Debye length, determined by the temperature and number density of the charged particles, there can be charge imbalance.

If a negative test charge is introduced, the electrons are repelled and the ions are attracted. Very quickly, the resulting displacement of the electrons and ions produces a polarization charge that acts to shield the plasma from the test charge, an effect known as Debye shielding. The characteristic length over which a charged particle's bare electric field has substantial influence, i.e. the length over which shielding occurs, is called the Debye length, which is found to be

$$\lambda_D \equiv \left(\frac{\varepsilon_o kT}{n_o e^2} \right)^{1/2} \tag{2.1}$$

where ε_o is the permittivity of free space, kT the term describing the electron kinetic energy, n_o and e the number of electron density and the electronic charge, respectively. A simple practical formula for Debye length is

$$\lambda_D = 6.9 \left(\frac{T}{n_o} \right)^{1/2} cm \tag{2.2}$$

where T is in K and n_o is in cm^{-3}. It is a rough measure of the size of the Debye shielding cloud that the charged particle carries with itself. For distances exceeding the Debye length, the electric field of an individual charged particle is effectively shielded out by the surrounding plasma. The number of particles in the Debye sphere N_D is defined as the "plasma parameter",

$$N_D \equiv n_o \cdot \frac{4}{3} \pi \lambda_D{}^3 \tag{2.3}$$

If $N_D < \approx 1$, collective phenomena are of little importance and this is known as an "independent particle". This means that collisions dominate. If $N_D >> 1$, a huge number of particles are collectively responsible for the Debye cloud, and that is called "collective behavior". Thus, collective effects dominate over collisions, which is the criterion of plasma.

Since plasma particles behave collectively, the plasma can support a wide variety of wave motions and oscillations. If electrons in uniform and homogenous plasmas are displaced from their equilibrium position, an electric field arises because of charge separation. This electric field produces a restoring force on the displaced electrons. Since the magnitude of the charge imbalance is directly proportional to the displacement, the restoring force is given by Hooke's law, $F = -k\Delta x$ where, Δx is the displacement and k is the spring constant. Since the electrons have inertia, the system behaves as a harmonic oscillator and the electrons overshoot and oscillate around the equilibrium position. It is known as plasma frequency and expressed by

$$\omega_D = \left(\frac{n_o e^2}{\varepsilon_o kT} \right)^{1/2} \tag{2.4}$$

It is mentioned that plasma will shield out an electric field at frequencies below the plasma frequency, because the electrons can move to shield the charge or the potential imbalance induced by the applied field. Based on the relative temperature of the electrons, ions, and neutrals, plasmas are classified as thermal or non-thermal. In most cases the electrons are close enough to thermal equilibrium that their temperature is well-defined. The temperature of particles is directly proportional to their average random kinetic energy. In the thermal equilibrium the distribution of velocities of particles of type s is given by the Maxwellian distribution:

$$f_s(v) = n_s \left(\frac{m_s}{2\pi kT_s} \right)^{3/2} e^{-\frac{mv^2{}_s}{2kT_s}} \tag{2.5}$$

where n_s the number of density of each species is, $f_s(v)$ is the distribution function, v is the velocity, m_s is the mass of the particles, k

is the Boltzmann's constant, and T_s is the temperature. Thus, using this distribution the average kinetic energy can be obtained as

$$E_{av} = \left\langle \frac{1}{2} m_s v^2 \right\rangle = \frac{3}{2} kT \quad (2.6)$$

The equation above shows that the temperature is directly proportional to the kinetic energy of the particles. For plasma at thermal equilibrium, the distribution function for each species should be Maxwellian and the temperature of all species must be equal. Thus, thermal plasma is produced at high pressures (>10 kPa) with electron and ion temperatures of the order of 1-2 eV and with very low gas ionization (also known as atmospheric pressure plasmas). Due to high gas temperature, atmospheric plasmas such as corona may be preferred for surface etching, but are less suitable for thermally sensitive material processing. Moreover, atmospheric plasmas have certain disadvantages as compared with low pressure plasmas (LPD): i) it is difficult to sustain a steady discharge, ii) higher voltages are required for gas breakdown iii) it is difficult to form uniform plasma throughout the reactor volume.[7]
In addition, because collisions occur very infrequently in tenuous plasma, "low pressure plasma", the approach to thermal equilibrium is very slow. Therefore, non-equilibrium effects are quite common in plasma "glow discharge". Since the electron and ion masses are very different, the rate of energy transfer is much slower than between electrons or between ions. Therefore, when the plasma is heated, a substantial temperature difference is often developed between electrons and ions. Thus, for glow discharges, ions and neutral have much lower temperatures, whereas the electrons are much hotter. As Verprek[8] points out, due to the highly non-equilibrium nature of low pressure glows, these systems are capable of "high-temperature chemistry at low temperature".[9] Most glow discharges, in other words, cold plasmas, are obtained when the gas pressure is in the range of 0.1 to 10^3 Pa. These low pressures allow for a relatively long mean free path of accelerated electrons and ions. The ions and neutral particles are near ambient temperatures and the free path of the electrons which are at high temperature or electron-volts level is long. Therefore, the electrons undergo relatively low-impact collisions with molecules at this pressure, and the reaction remains at low temperatures. LPD or glow discharges are the form of plasma processing implemented for the surface modifications described in this dissertation. Hence, LPD will be the focus for the remainder of this chapter.

2.2. Plasma Generation

The ionization of gases is accomplished by applying an electrical field in the form of direct current (DC), low frequency (LF<100 KHz), radio frequency (RF 2-100 MHz, e.g. 13.56 MHz), or microwave frequency (MW 2.45 GHz).
A DC glow discharge is produced by applying a DC voltage to two conductive electrodes that are inserted into a gas at low pressure. A high-impedance power supply is used to provide the electrical field. Samples placed on the cathode of the DC discharge are exposed to bombardment by high-energy ions that are accelerated at voltages that must be above the minimal breakdown voltage. This might cause damage to sensitive substrates.
Low frequency is the least expensive method of energy field generation. Unfortunately, low frequency is also the least efficient method for surface modification. The efficiency of the reaction is related to the energy necessary to

sustain ionization, the intensity and the frequency of vacuum ultraviolet (VUV) radiation.
In the case of MW excitation, no electrodes are necessary. MW source plasmas are generated downstream or in a secondary environment. Downstream is defined as the plasma generated in one chamber and drawn by a vacuum differential into the work area or another chamber. This can be advantageous for etching or organic removal and effective for ion-sensitive components due to the higher degree of ionization in general and especially for MW-ECR (Electron Cyclotron Resonance) discharges. However the use of the ECR effect is restricted to the pressure range <1 Pa. MW discharges are also prone to substrate heating problems and they also produce a less homogenous process resulting in the compromising of uniformity across the work area. In surface modification the effective depth of the modification is tens of nanometers, so the uniformity of the process becomes increasingly important, rendering MW source plasmas a less desirable choice. MW plasmas are either smaller, highly localized discharges, or thermal plasmas which are less suitable for textile material processing applications. Moreover, a set up with MW excitation is more complicated and more expensive compared to an RF setup.

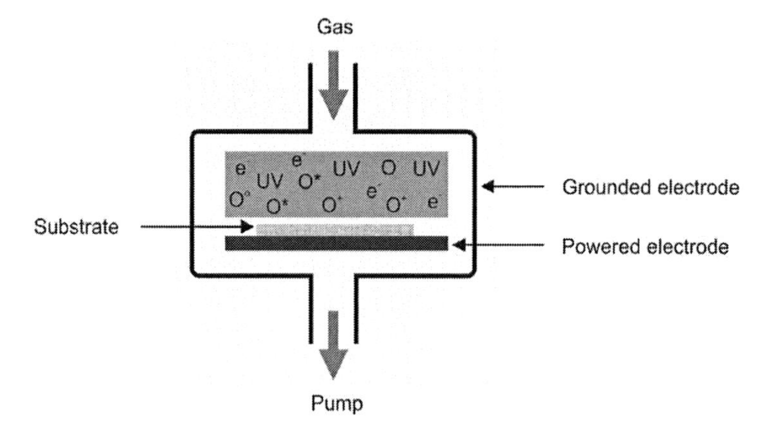

Figure 1. RF plasma set up and illustration of active species present in oxygen plasma.

The setup for RF excitation is the simplest one. In the case of a capacitively coupled RF discharge, two electrodes are mounted in a vacuum chamber as shown in Figure 1. A process gas with a typical pressure of a few pascals is introduced. When the RF voltage exceeds a certain value in the range of some hundred volts, depending on gas, pressure and reactor geometry, the discharge ignites. RF plasmas exhibit significantly higher levels of VUV, which in part explains the concentrations of electronically charged particles are higher than those found in other plasma sources. RF plasmas have also been noted to be more homogenous, a trait that is critical in treating irregularly shaped and overly large objects. RF plasmas are characterized by higher ionization efficiencies and can be sustained at lower gas pressures than DC discharges. Finally, in the case of RF discharge, the energy of the ions bombarding the sample is controlled by the negative bias, which can be adjusted over a wide range of values. In the scope of this dissertation, LPD RF discharges, so-called low temperature plasma, were performed. Their characteristics will be explained as follows.

2.3. Plasma-Surface Interaction

Interaction mechanisms between plasma and a polymer surface are very complex, for they include physical bombardment by energetic particles and by VUV photons and chemical reactions at the surface. Four main effects result out of this, surface cleaning, ablation, crosslinking, and surface chemical modification, as can be seen in Figure 2. These occur together in a complex synergy, which depends on many parameters. Nevertheless, it is possible to control the main set of parameters governing a given process, and thereby to assure reliable reproducibility of the process outcome, a factor quite important for industrial implementation.
The ion bombardment of the surface causes sputtering effects. If the substrate is properly biased (by applying a negative potential), the energy of the ions can be greatly increased yielding higher sputter rates. Furthermore, ions can chemically react with the surface. Thus, the combined action of chemically reactive and accelerated ions results in anisotropical etching of substrate surface, which is known as reactive ion etching. The VUV radiation with wavelength of less than 178 nm can cause photoionization.[10] Furthermore, VUV radiation can cause dissociation of bonds yielding free radicals.[11] This can lead to chain scission, rearrangements or elimination of specific functional groups.[12] The radicals created on the surface can cause crosslinking,[13] and can react with species from the plasma phase or subsequently can react with oxygen when the surface is exposed to atmosphere. Furthermore, bombardment of electrons on the surface seldom causes chemical effects. Because low energetic electrons have a substantial penetration depth (larger than 500 Å for 20 eV electrons)[14], their action is not restricted to the outermost surface. It seems likely that the electrons are substantially retarded when they approach a self bias substrate in plasma, resulting in an increased penetration depth of accelerated ions.
The radicals colliding with the surface of the substrate can be incorporated at the surface (radical-radical recombination), can abstract hydrogen or other atoms from the surface, or can induce polymerization and/or crosslinking of the surface. The effect of the bombardment of neutral species on the surface is largely dependent on their chemical reactivity. For example, unsaturated species can polymerize with a radical containing surface, but e.g. argon atoms will not react with such a surface. The overall effect of the plasma process is determined by the gas used during the discharge and to a lesser extent by the conditions.

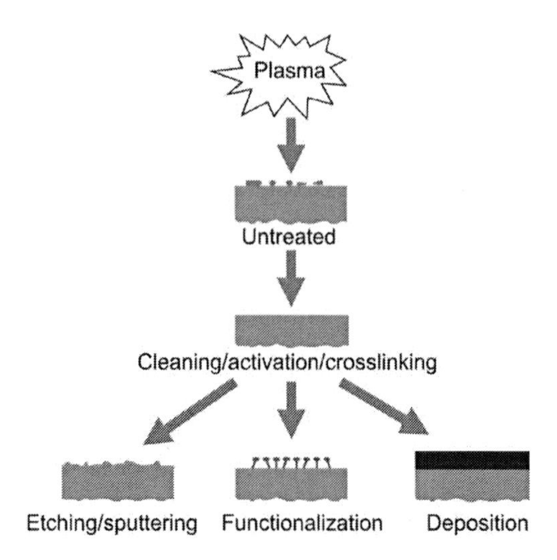

Figure 2. Schematic diagram of the interaction of plasmas with solid surfaces.

2.4. Plasma Processing

LPD can be used for surface modifications ranging from simple topographical changes to the creation of surface chemistries and coatings that are radically different with respect to their bulk materials.[15] Based on a wide range of applications and mechanism involved in plasma technology, plasma processes which are environmentally safe methods can be subdivided into two main categories:

- plasma modification referring to surface cleaning, activation, and surface etching, and
- plasma polymerization referring to surface chemistry restructuring via deposition.

2.4.1. Plasma Surface Modification

Gas at low pressure allows the acceleration of free electrons when driven by an external source (RF generator). As a consequence, highly reactive and activated molecular species such as chemical radicals, ions, electrons etc. can be created by ionization, fragmentation (dissociation), excitation, UV radiation, etching reactions etc. These species chemically and physically react with the polymer surfaces thus altering the surface properties and surface morphology in the topmost layers (a few nanometers). Non-polymer forming inorganic gases are used in this plasma activation process.

Surface activation changes from surface cleaning, radical formation and atom implantation to surface etching; it depends on different process parameters such as for example energy input. Surface cleaning is commonly used prior to other processing steps such as polymerization, metallization, dyeing, lamination etc. in

order to increase the adhesion between plasma coating and substrate surfaces (for example metals, polymers, textiles etc.). An oxygen containing gaseous mixture (Ar/He with O_2) was found to be more efficient compared to pure inert gases (Ar, He etc.) to remove organic contaminants by oxidizing polymer surfaces and producing plasma degradation products such as hydrogen, steam, carbon-dioxide etc.[27]
A surface functionalization of the polymer surface can be carried out by the incorporation of radicals or atoms. The modification can be altered over a wide range of properties from hydrophobic to hydrophilic, depending on the plasma power and gas feed. The use of gaseous mixtures facilitates crosslinking. For instance, the addition of argon produces better hydrophilicity and wettability, as compared to pure oxygen plasma.[20] Furthermore, surface functionalities enhance the polarity and the adhesion at the interface for a second layer such as a plasma polymer film when it contains the respective reactive groups. On the other hand, radical generation on the polymer surface is used to enhance chemical bonds with plasma polymers and increase adhesion of the plasma coatings. Plasma activation at moderate power generates radicals mainly by hydrogen abstraction from the polymer chain during collisions of electrons with the polymer surfaces. Electrons, UV radiation or ion bombardment can generate radicals by C-C bond scission of the polymer.
Plasma cleaning of polymers and metals is a well-known and effective process to remove contaminations such as sizes, additives, anti-oxidants, dust, process residues, carbon residues, oil, or other organic compounds.[16-18] The presence of contaminants on the surface prevents adequate adhesion with the subsequent processes such as metallization, lamination, plasma coatings etc.[19] Whereas wet chemical treatments leave nano-meter thick contaminant-layers, LPD was found to be a very efficient method of surface modification, even as compared to MW and atmospheric plasmas.[20] Oxygen and/or an inert gas are commonly used as a source of plasma gas. In this process, oxygen plasma causes reactions with surface contaminants resulting in their volatilization and removal of the degradation products as water vapor, CO, CO_2, H_2, etc. Moreover, oxygenated functions such as –OH, -C=O, -COO, -COOR etc. are grafted onto the surface. Thus, functionalized and activated surfaces can be obtained, which can be used to improve wettability,[21,22] dyeability, printability, shrink resistance of wool[23] in textiles.[24] The grafted surfaces can also lead to covalent bonds suitable for further attachment of biomolecules, coatings, matrices etc.[25] These new reactive sites can be used to improve adhesion and prevent delamination of the subsequent coatings, thus supporting the formation of abrasion resistant coatings. In the case of composites, the hydrophilic polar surfaces enhance chemical bonding to thermoset matrix polymers such as epoxies etc. Improvement of the interfacial shear strength in fiber/matrix composites can increase the strength of the composites. Another major effect can be achieved by inert gas plasma action, i.e. surface crosslinking, commonly known as CASING process, as described by Arefi-Khonsari 2005 et al.[26] The plasma is assumed to break C-C, C-H bonds by ion and VUV bombardment. These free radicals in turn recombine on the surface causing a stable crosslinking of the surface structure. It can cohesively strengthen the topmost surface layers by creating new bonds, such as C=C, which indicates crosslinking and produces a stronger and harder substrate micro-surface.
At high plasma energy levels, etching and texturing occur by removal of polymer from the surface. Rough and textured surfaces can be achieved by physical and chemical interaction of reactive species generated in the plasma zone. The rate of etching (a few nanometers per second) strongly depends on the substrate materials

which are being etched and the energy level of the plasma. On the other hand, it relies on the ion bombardment, radicals and ion density, UV radiation and plasma gas. In addition, it has a strong influence on surface energy and thus wettability. One mechanism by which a surface is ablated is known as surface etching or texturing. Roughening of the surface, which is usually on the nanometer to micrometer range, can play a significant part in adhesion by increasing the total contact area between the adhesive and the subsurface. As a result, micro-bonding can be enhanced greatly in reinforced composites.[27,28] The breakdown of weak covalent bonds in a polymer results in removal of the outermost molecular layers through bombardment with high energetic particles (i.e. atomic species), which causes small volatile fragments to evaporate. Ablation of materials by plasma can occur as part of two principle processes: physical sputtering and chemical etching. The physical sputtering of materials is accomplished by chemically non-reactive plasmas such as inert gas plasma (Ar, He etc.). Chemical etching occurs in chemically reactive types of plasmas using organic and inorganic gases, such as oxygen, nitrogen, ammonia, fluorocarbon (CF_4) etc., which are chemically reactive and do not deposit polymers in their pure gas plasmas.

Plasma etches material in the direction perpendicular to the surface more rapidly than in the direction parallel to the surface. It has emerged as an important tool for microelectronic manufacturing for patterning, doping of multilayers, and creating conducting films, etc. in order to form and connect circuit elements like transistors and capacitors.[29] The most common strategy used to pattern thin films is lithography. The film is coated with a photoresist, which is a usually polymeric material that is light-sensitive. The photoresist is exposed to UV light through a mask that allows only certain areas of the coating to be illuminated and thus, the exposed portion of the coating will be removed.

2.4.2. Plasma Polymerization

In plasma deposition, which is commonly known as plasma polymerization or plasma-enhanced chemical vapor deposition (PE-CVD), a very thin polymer layer (nm to μm) is deposited on the substrate surface. The layer is formed through polymerization of a monomer, which is directly polymerized on the surface activated by plasma modification, as already discussed in the previous section. In contrast to classic polymerization, plasma polymerization can use every monomer gas or vapor which is not limited to their reactivity. Bradley et al. reported that plasma polymerization could be performed for almost any kind of monomer and it is mainly the elemental composition of the monomer, which is fed into the reaction that is important. The growth rate varied depending on the monomer structure even if polymerized films showed similar characteristics.[30]

Commonly used hydrocarbon gases are methane, acetylene (ethyne), and ethylene (ethene) etc., where the polymerization occurs either by abstraction of hydrogen atom from their starting molecules and/or by bond opening for unsaturated hydrocarbons such as ethylene etc. These gases can also be mixed with inorganic gases such as ammonia, oxygen, water vapor, carbon dioxide, nitrogen etc. depending on the desired surface properties. The treatment of hydrophilic substrate-surface with organic compounds, such as hexamethyldisiloxane (HMDSO) or fluorocarbon gas (C_3F_6, CF_4 etc.), can change surface properties and results in a decrease of water absorption or wettability.[31,32] Thus, the materials become highly

hydrophobic with high of contact angles (CAs). The polymerization of HMDSO has been proven to form scratch resistant, transparent layers containing Si-C and Si-O-Si functionalities. As a result, the new structures are found to be chemically inert and crosslinked in nature as described by Tajima 1985.[33] By addition of oxygen they can form glass-like coatings. McCord et al.[34] compared plasma polymerization of CF_4 and C_3F_6 to improve hydrophobicity of cotton fabrics. They found that both plasmas generated fluorocarbons groups (-CF, -CF_2, and –CF_3), while C_3F_6 polymerization generates more –CF_3 groups, highly related to hydrophobicity, than CF_4 plasma.
In plasma polymerization, the monomer is fragmented under plasma conditions and builds up a plasma polymer. The plasma polymer does not contain regular repeating units; the chains are branched and randomly terminated with a high degree of crosslinking. Thus, it has highly crosslinked and disordered structure without repeating units, as shown in Figure 3. Structural preservation and gradients, with increasing degree of crosslinking over film thickness, can be controlled through process parameters, such as gas pressure, gas flow, and applied electric voltage, so that one can also construct so-called gradient layers. It is very easy to obtain ultra-thin films[35] with very useful properties for technological applications. A combination of polymerizable gases with non-polymerizable gases allows for the deposition of a variety of plasma polymer layers with many different functional groups possible. Thus, depending on the selection of the gas, monomer, process parameters, these thin coatings can be deposited with various physical and chemical characteristics. Consequently, functionalized surfaces with special properties can be obtained, which is one of the principal advantages of plasma polymerization. In addition, since the plasma treatment is largely independent of the substrate material, this leads to the possibility to use a universal coating process instead of optimizing surface modification processes and plasma parameters for each different substrate material. However, special care should be taken in particular for textile substrates due to their 3-D structure and manufacturing residuals.

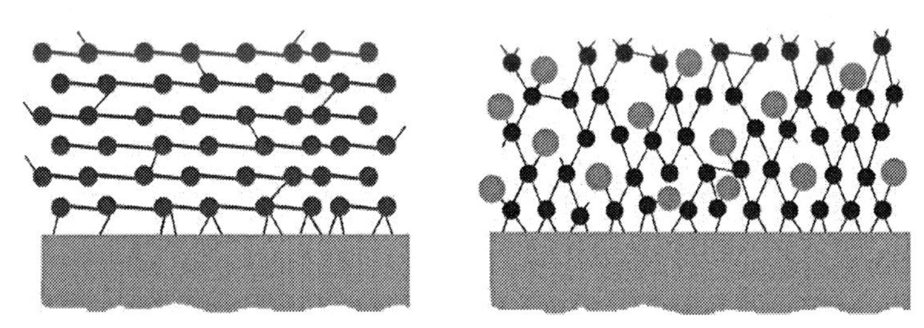

Figure 3. The illustration of conventional polymer (left) and crosslinked plasma polymer (right).

2.4.3. Mechanism of Plasma Polymerization

Since plasma is inherently a reactive gaseous mixture, many different reactions occur simultaneously in a glow discharge. A scheme of interaction between a solid phase and a plasma is summarized in Figure 4.[36] More specifically, there are two major processes having opposite effects, deposition and ablation, as described previously. Besides the discharge conditions, such as energy density, the plasma gas mainly determines which of the two processes is dominant.[37] The plasma ablation competes with polymer formation in almost all cases when plasma is used

to treat surfaces of solid materials. Since radical processes dominate the macromolecule formation yielding mainly amorphous and crosslinked structures, plasma polymerization is commonly described as radical-dominated polymerization.
The growth mechanism of polymerization is known as rapid step-growth polymerization (RSGP).[6] The overall reactions contains two major routes of RSGP as described by Yasuda. Cycle 1 and Cycle 2 consist of reaction species from mono functional and difunctional reactive species, respectively. Cross-cycle reactions from 2 to 1 can occur. Since, the reactive species are mainly radicals, thus the recombination of reactive species and reactivation of reaction products determine the RSGP process. Furthermore, the surface takes part in plasma polymerization by radical sites, third-body reactions and etching processes that lead to ablation and re-deposition. Hence, a plasma polymer typically results from rivaling etching and deposition processes depending on the plasma species present during film growth yielding a more or less crosslinked structure.
It is assumed that the gas particles (monomers and reactive gases) travel through an active zone (bulk plasma and plasma/sheath boundary region), where radical formation is taking place within the gas phase, and then enter a passive zone (plasma sheath and surface growth region) yielding recombination and stable products such as deposition (concept of chemical quasi-equilibria).[38,39]

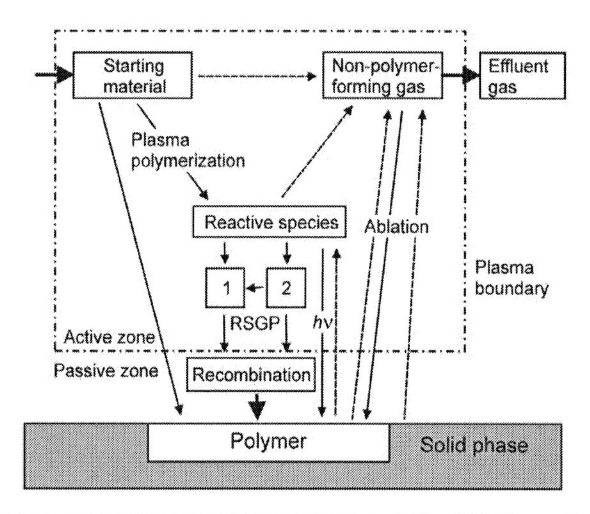

Figure 4. Schematic mechanism of rapid-step growth polymerization (RSGP) considering an active and a passive plasma zone.[4]

2.5. Plasma Technology in Textile Processing

The applications of textile fibers are directly related to their unique characteristics such as geometrical, physical, and chemical properties. Some examples of chemical properties of fibers include resistance to the effects of acids, alkalis, oxidizing agent etc. Most textile fibers are to some extent hygroscopic, and therefore they are capable of absorbing moisture from the atmosphere. This character is a direct reflection of their chemical nature. Textile fibers can vary from hydrophilic, such as cellulosic fibers (i.e. cotton, hemp, flax, ramie etc.), to those that are essentially hydrophobic, such as many synthetic fibers (i.e. polyester, acetate, polypropylene etc.). The rate and extent of the dye sorption, which is the affinity of various dyes for textile fibers, depend on the chemical structure of the dye and the fiber. The physical properties of textile fibers, and indeed of all materials, reflect their molecular structure and intermolecular organization. For instance, all polymeric fibers are semi-crystalline i.e. the polymeric chains are partially ordered into domains. Another important aspect of many fibers, such as carbon, aramid, glass, polyolefin fibers etc., is that they posses high strength, durability, and low density; thus, they have found increasing application as reinforcements in fiber-reinforced composites. While they already show excellent potential, tailoring the surface properties further advances their technical use in the modern world.

Plasma surface treatments have long been utilized to modify the chemical and physical structures of the surface layers of textile fibers, thus improving the properties of fibers in many technological applications. Atmospheric pressure plasmas such as corona or dielectric barrier discharges, however, might be of special interest for the textile industry due to an easier processibility.[40] During the past decade, considerable efforts have been made to generate stable atmospheric pressure plasmas.[41] While they show some use for activation and hydrophilization of textiles,[42-44] it is rather difficult to obtain high quality plasma-polymerized coatings on textiles with atmospheric techniques. Thus, LPD has been investigated for the enhancement and/or replacement of conventional wet-chemical processes in obtaining functional coatings in technical-textile applications.[45] First attempts to deposit plasma polymers on natural and synthetic fibers within a continuous LPD reactor, where the fibers are processed air-to-air, were performed in the early 1980s.[46] Phosphorus containing monomers were used to impart flame retardancy to cellulosic fibers and acrylamide plasmas to improve the physico-mechanical properties. Although, within the textile industry up to now merely special applications and niche products are performed with the help of plasma polymerization, LPD polymerization has significant potential to stimulate the textile sector. In recent years, plasma polymerization on textiles and fibers was mainly investigated for hydrophobization, permanent hydrophilization, dyeability, adhesion improvement in fiber reinforced composites,[24] and improvement of conductivity in e-textiles.

2.6. Characterization of Plasma-modified Surfaces

In order to characterize material surfaces a wide variety of methods are available nowadays. However, the techniques, available at Empa, and implemented in this work, are described here. For plasma modified materials a distinction between plasma treated and plasma polymerized materials has to be made. During plasma polymerization processes usually a polymer thickness of 20 nm or more is deposited. By a proper selection of the substrate and its area, a substantial amount of plasma polymer can be obtained. This enables analysis by some general solid state methods. Indirect methods such as CA measurements, which are sensitive to the outermost 5 Å, dyeing, and adhesion can be used for plasma treated films. However, in both cases non-coated and coated surfaces can mostly be utilized by the following analytical techniques such as thickness measurement, elemental analysis, topographical analysis, CA measurement, capillary rise test, abrasion test, and colorimetric analysis etc.

2.6.1. Thickness Measurement

The Profilometer is an easy and fast analytical tool, in particular for thickness measurements and ideal for R&D and industrial quality processes. The region of interest can be focused easily by optical microscope. The very thin (2.5 nm – 400 μm) step heights can be determined precisely as long as suitable edge on the coated film can be achieved. The surface topography like surface microroughness and waviness in the micrometer range is also possible. The process has excellent repeatability and reproducibility. It offers comprehensive process control and plotting features (HRP-75, KLA Tencor), as shown in Figure 5.

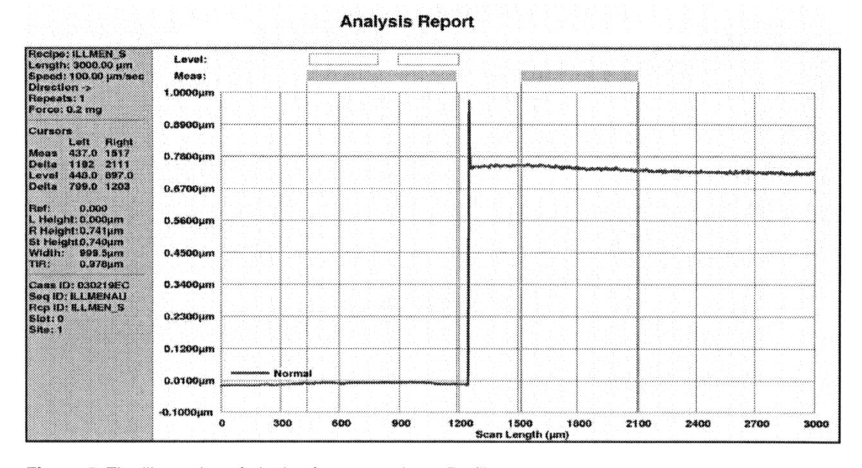

Figure 5. The illustration of plotting features using a Profilometer.

2.6.2. Elemental Analysis

The functional groups at the polymer surface can be analyzed qualitatively and quantitatively by XPS and IR. XPS measurements were performed using a PHI LS 5600 instrument with standard Mgkα X-ray source. The energy resolution of the spectrometer was set at 0.80 eV/step at a pass energy of 187.85 eV for survey scans and 0.25 eV/step and 58.70 eV pass energy for regions scans. The X-ray beam was operated at a current of 25 mA and an acceleration voltage of 13 kV. Charge effects were corrected using Carbon 1s = 285.0 eV. The concentrations of the surface species were determined by CasaXP software (peak areas using the instrument specific relative sensitivity factors).

In a vacuum the substrate surface is irradiated with X-ray photons and they interact with an inner-shell electron of an atom. The energy of an X-ray photon is transferred to the electron and it received enough energy to leave the atom and escape from the surface of the substrate. Thus, a photoelectron with certain kinetic energy is created. As displayed in Figure 6, photoemission can occur after the photon transfers its energy to a core-level electron leading to the emission of a photoelectron from the n-electron initial state. Reorganization of the electron distribution of an atom may occur following photoelectron emission and manifests itself by an electron dropping from a higher energy level to the vacant core hole. Once the vacant core has been filled, the atom can rid itself of the excess energy by emitting an electron from a higher energy shell, known as an Auger electron. If an Auger electron is not emitted, the atom may choose to rid itself of the excess energy in another manner, X-ray emission, through a process known as X-ray fluorescence. Thus, the XPS wide scan, or survey spectrum does not only display evidence of photoelectrons, but may also contain Auger electrons. However, Auger peaks are not studied in the scope of this dissertation, and the focus will be on the photoelectrons.

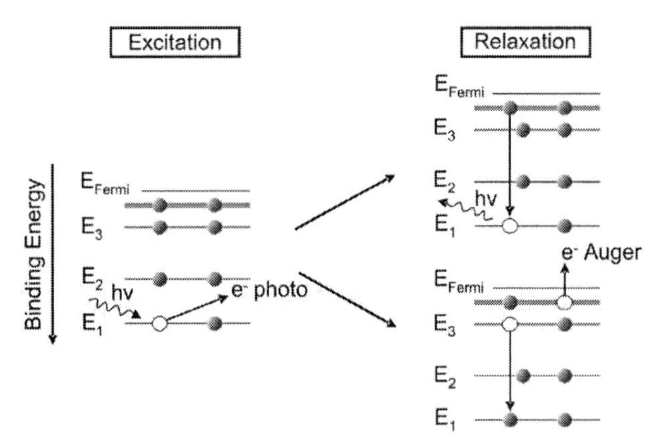

Figure 6. Schematic diagram of photoelectron and Auger electron emissions.[47]

Since the kinetic energy (E_K) of the photoelectron is measurable, therefore the binding energy (E_b) of the electron can be calculated:

$$E_b = h\nu - E_K \tag{2.7}$$

where $h\nu$ is the incident X-ray energy (eV), h is the Plank's constant, ν is the X-ray frequency and E_K is the kinetic energy of the emitted photoelectron (eV). Each element has a unique binding energy, which can be used to identify the elemental composition of the substrate. The photoelectrons are separated according to their kinetic energy and the number of photoelectrons with different energies is counted. The energy of the photoelectrons is related to the atomic and molecular environment from which they originated and the number of electrons emitted is related to the concentration of the emitting atom in the substrate.

In addition, elemental depth profiles, several hundred nanometers into the sample, provide the in-depth compositional analysis, which is commonly done by Ar^+ ion beam sputtering. Surface sputtering allows depth profiling in order to determine the composition of a layer and to remove possible surface contaminations. However, it can influence the chemical composition by preferential sputtering. Generally different oxidation states of one element can be clearly distinguished by XPS, also known as ESCA (Electron Spectroscopy for Chemical Analysis), whereas the separation of e.g., oxygen containing functional groups with the same oxidation state (e.g. ethers, hydroxyl, etc.) is not possible. It is mentioned that the same atoms in different binding states give rise to distinct differences in binding energies. Table 1 reports the binding energies of some functional groups identified in the scope of this dissertation.

Table 1. Typical binding energies for C 1s, N 1s and O 1s

Functional Group		Binding Energy (± 0.2 eV)
C 1s		
Hydrocarbon	C-C/C-H	285.0
Amine	C-N, C=N	286.3
Alcohol/Ether	C-O-H, C-O-C	286.4
Ester	C-O-C=O	286.9
Carbonyl	C=O	287.6
Acid/Ester	O-C=O	288.9
Amide	N-C=O	288.2
N 1s		
C bound to N	N-C	398.5
C bound to N	N=C	399.6
O bound to N	N-O	400.6
O 1s		
Carbonyl	O=C /O-C=O	532.2
Alcohol/Ether	C-O-H, C-O-C	532.6
Acid/Ester	C-O-C=O	533.6

2.6.3. Surface Topographical Analysis

AFM Analysis

The topographical changes of plasma-coated surfaces can be effectively characterized by Atomic Force Microscopy (AFM) (Solver Pro, NT-MDT, Russia) in contact, non-contact or tapping mode operations. The machine includes a very fine tip with a nanometer tip-radius to scan surfaces to be examined, which is mounted on a micro-machined cantilever. When the tip scans a surface, inter-atomic forces between the tip and the sample surface induce displacement of the tip and corresponding bending of the cantilever. A simple beam deflection system is used to monitor the cantilever displacement and measure cantilever deflection with a sub-angstrom resolution and its twisting angle. A laser beam is focused on the back side of the cantilever close to the tip position. The reflected laser beam is then fallen onto the four quadrant photo diode. The signal regenerates on the piezo crystal and the tip is controlled by the piezo crystal. The movement of the piezos is recorded as topographical picture by a computer. A typical system used for AFM analysis is illustrated in Figure 7.

Figure 8 also demonstrates that there are 3 regimes present, as a function of tip-to-sample separation distance, including contact, intermittent contact or tapping, and non-contact regimes, which are in fact different modes of AFM analysis. For each of these regimes, the total attractive and repulsive forces differ.

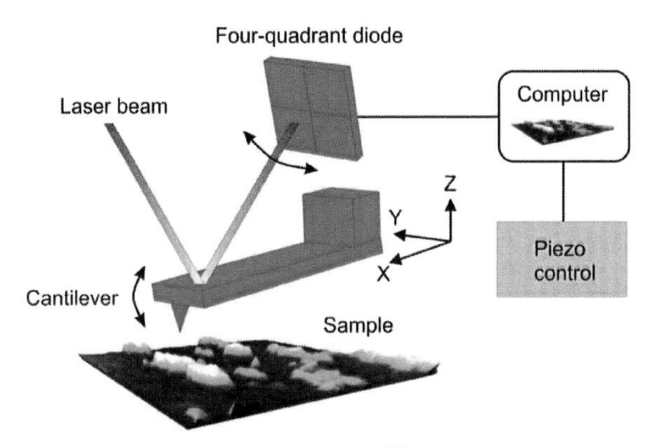

Figure 7. Schematic diagram of a typical AFM apparatus.[48]

In the contact mode regime, the AFM tip makes soft physical contact with the surface, and the force, F, is of a repulsive nature. The tip is gently scanned over the sample, and the contact forces the cantilever to bend to accommodate the changes in topography of the surface. In this manner a topography map can be generated.

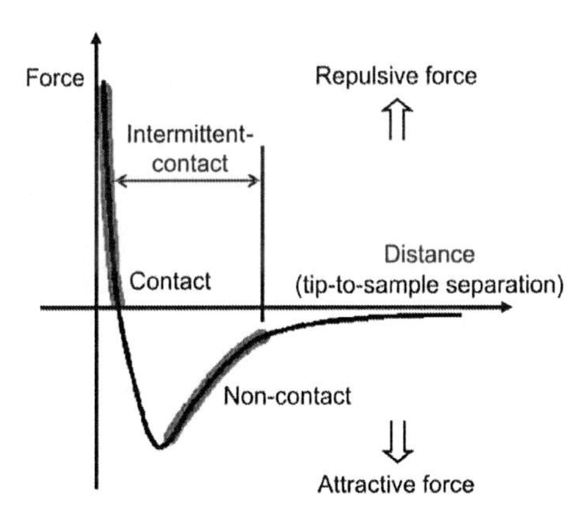

Figure 8. Schematic representation of the interatomic force as a function of distance.[49]

In the non-contact mode regime, the tip-to-sample distance is of the order of 10-100 nm. The force, F, is attractive and is a result of long-range van der Waals interactions. Far away from the sample, the cantilever is vibrated at a frequency just above the resonance frequency ω_o. When the tip is brought closer to the sample, the resonance frequency ω_r will decrease according to equation

$$\omega_r = \sqrt{\omega_o^2 - (1/m)\delta F / \delta z} \qquad (2.8)$$

This induces a decrease of the amplitude of the vibration. The amplitude is then used to control tip-sample distance.

The intermittent contact or tapping mode regime relies on the same principles as the NCM imaging, but the cantilever is now driven at a frequency slightly below the resonance frequency. The amplitude of the vibration will increase when the tip is brought closer to the sample, until the point at which the tip touches the sample. This induces a decrease of the vibration amplitude, which is used to control the tip-sample distance.

SEM Analysis

The surface of an object can be scanned with an ultrafine electron beam (d $\approx$5 nm) using Scanning Electron Microscope (SEM) (Hitachi, S-4800). The SEM's principle is a focused beam of electrons passing through a vacuum and making contact with a solid of interest. Those electrons bombard the sample and are released as secondary electrons (SE) or backscattered electrons (BE). Low energy SE are produced by the inelastic interactions between the sample and the incident electron beam. High energy BE are produced by elastic interactions between the sample and the incident electron beam. The SEM's primary imaging method is through the collection of those electrons that are released by the sample. These electrons are detected by a scintillation material, which is a radiation detector that produces flashes of light from the electrons. The light flashes are then detected and amplified by a photomultiplier tube. SE imaging provides high resolution imaging of surfaces (for ideal sample $\approx$3.5 nm). On the other hand, since BE can escape from much deeper than SE, BE surface topography is not as accurately resolved (optimum resolution $\approx$5.5 nm).

2.6.4. Contact Angle Measurement

Contact angle (CA) measurement is an easy to perform method for surface analysis related to surface energy and surface tension of liquid droplets at solid interfaces. The CA describes the shape of a liquid droplet resting on a solid surface. CA is the angle between the outline tangent of the drop at the contact location and the solid surface. This technique is extremely surface sensitive, with the ability to detect properties on monolayers related to wetting, adhesion, and absorption.

When a droplet of liquid rests on the surface of a solid, the shape of the droplet is determined by the balance of the interfacial liquid/vapor/solid forces (Figure 9). The theoretical description of CA arises from the consideration of a thermodynamic equilibrium of the above three phases. At equilibrium the Young's equation is given by

$$0 = \gamma_{SV} - \gamma_{SL} - \gamma_{LV}\cos\theta \qquad (2.9)$$

Where γ, is the liquid-vapor energy (surface tension) and the indices L, V, and S represents the liquid, vapor and solid phases, respectively, and θ refers to experimental (equilibrium) CA. The left term is the projection of the vector γ_{LV} on the planar surface. It also illustrates that γ_{SV} is directly dependent on the surface energy of the solid, γ_S and the vapor spreading pressure, π_e.[50]

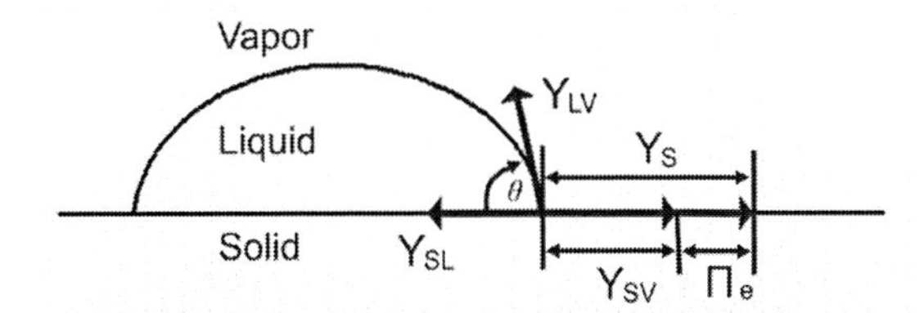

Figure 9. Schematic representation of Young's equation, illustrating a sessile drop of liquid on a solid surface.

Thus, if a liquid with well-known properties is used, the resulting interfacial tension can be used to identify the nature i.e. interfacial energy of the solid. If the water CA is less than 90°, the surface exhibits hydrophilic and if larger than 90° exhibits hydrophobic nature. Superhydrophilic surfaces show a CA less than 10°, whereas a superhydrophobic surface shows a CA at about 140° or more.
In this work, changes in surface hydrophilization in terms of CAs induced by plasma treatments were measured by the optical inspection method using G10 (Krüss GmbH, Germany) (Figure 10). Before the measurement all samples were conditioned at ambient temperature and the static CAs were measured with drop size of $\approx 10\ \mu$L with de-ionized water on coated glass substrates as well as on the coated textiles in a conditioned room (20°C and 65% RH). For each sample three measurements were taken at different locations and the average of these measurements were reported.

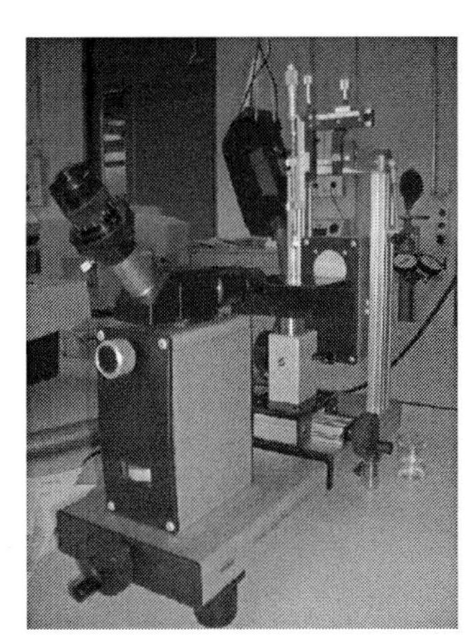

Figure 10. Used contact angle apparatus.

2.6.5. Abrasion Test (Martindale)

The Martindale is a well known test to determine the abrasion and pilling resistance of all kinds of textiles, and leather. The Martindale abrasion tester (James Heel Nu-Martindale, England) used in this work is shown in Figure 11. The abrasion

resistance of plasma thin films on textile substrates can be measured which helps to know the degree of wear performance and durability of the coatings.
In this method, samples are rubbed against known abradents at low pressure (load) (commonly at 12 kPa) in continuously changing directions (Lissajou Figures) and the amount of abrasion or pilling is compared against standard parameters. Resistance to abrasion is evaluated by various means, including comparison to visual aids in the form of photographs, or actual samples. In addition, after the abrasion test the substrates can also be effectively evaluated by SEM and even by CA measurements with various liquids such as water, oils etc. The unique design allows removal of individual sample holders (6 holders) without lifting the top motion plate.

Figure 11. Martindale abrasion resistance tester.

2.6.6. Dyeing and Colorimetric Analysis

Dyes, organic compounds which are soluble in appropriate solvents such as water, are absorbed in solution on the substrate surface and diffuse into the interior material under certain conditions such as temperature, dyebath pH, dye concentration, and liquor ratio etc. Then, they can chemically bind with the functional groups of the substrate material. Thus, dyeing can be used to trace chemical groups present in the plasma polymers. The level dyeing can characterize the uniform distribution of functionalities in the thin films. Color measurement is used to measure color intensity of plasma-coated and dyed substrates.
The CIE color system is the most widely used method in color measurement which is based on the spectral power distribution of light emitted from a colored object and is factored by sensibility curves which have been measured for the human eye. The CIELAB color space can be visualized as a three dimensional space, where every color in the space is determined by its color coordinates:[51]

- L^*- the lightness coordinate.
- a^*- the red/green coordinate, with $+a^*$ indicating red, and $-a^*$ indicating green.
- b^*- the yellow/blue coordinate, with $+b^*$ indicating yellow, and $-b^*$ indicating blue.

These color coordinates of an object are calculated as follows: the object is measured by a spectrophotometer, a light source (illuminant) and an observer (2° or

10°) is selected, trimulus values (X, Y, Z) are computed from the light-object-observer data and L^*, a^*, and b^* are transformed (computed) from the X, Y, Z data, using the CIE 1976 equations. The CIE coordinate axis define the three dimensional CIE color space.
Colors can also be described and located in CIELAB color space using an alternate method, that of specifying their L^*, C^*, and h^o coordinates. Where, L^* are the same coordinates as in $L^*a^*b^*$ and C^* (chroma) and h^o (hue angle) coordinates are computed from the a^* and b^* coordinates (Figure 12).

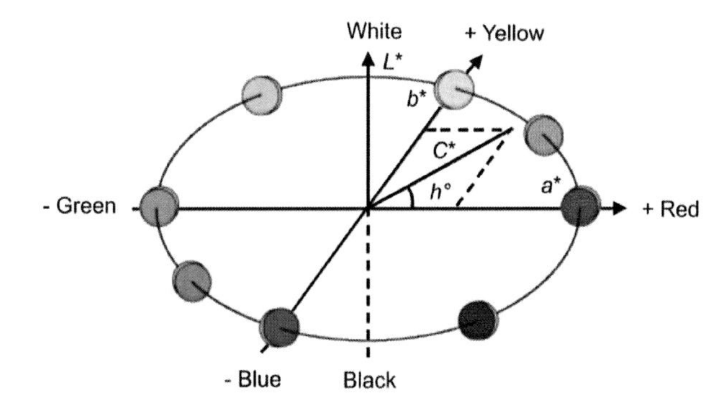

Figure 12. CIELAB color system.

In addition, the CIE color difference, between two colors in CIE space, is the distance between the color locations: DL^* is the lightness difference, Da^* is the red/green difference and Db^* is the yellow/blue difference. Thus, total color difference can be calculated as follows:

$$DE^* = \sqrt{\{(DL^*)^2 + (Da^*)^2 + (Db^*)^2\}} \quad (2.10)$$

In the scope of this dissertation, the color coordinators (L^*, a^*, b^*, C^* and h^o) were measured using a Datacolor Spectraflash (Datacolor AG, Switzerland). For each sample four measurements were performed and the average value was taken. The average deviation for each measurement was found to be very low. The reflectance (R%) value of the dyed fabrics was measured over the wavelength range of 360-750 nm. The illuminant type was D65 and the observer angle was 10°. The color strength (K/S value at 490 nm, where absorption coefficient is defined as "K" and scattering coefficient is defined as "S") of the fabrics were calculated from Kublka-Munk equation:

$$K/S = (1 - R)^2/2R \quad (2.11)$$

2.6.7. Tensile Strength Test

The ability of a material to resist breaking under tensile stress is one of the important and widely measured properties in the materials science industry. The tensile tests measure the force required to break the sample is known as "breaking strength" and the extent to which the sample stretches or elongates to that breaking point is known as “elongation”. The Instron 4502 (Serial Nr. H 3167, England), as shown in Figure 13, was used to measure strength and elongation according to SN 198 461 (1985). The experimental conditions are stated bellow:

- Room temperature: 20°C and 65% RH
- Stripe width: 50 mm
- Gauge length: 50 mm
- Speed: 25 mm/min

The machine is applicable for textiles, plastics, and felted materials etc. The sample is placed between grips of the machine at a specified grip separation and pulled until failure. The testing speed is determined by material specification. Tensile tests produce a stress-strain diagram, which is used to determine tensile modulus. Thus, a material's mechanical performance can be characterized.

Figure 13. Used tensile strength tester.

2.7. References

[1] L. Tonks, I. Langmuir, *Phys. Rev.* **1929**, *34*, 876.
[2] W. R. Gombotz, A. S. Hoffman, *Crit. Rev. Biocompat.* **1987**, *4*, 1.
[3] A. Grill, *Cold Plasma in Materials Fabrication: From Fundamentals to Applications*, IEEE Press, Piscataway, NJ, **1994**.
[4] H. Yasuda, "*Luminous Chemical Vapor Deposition and Interface Engineering*", Marcel Dekker, USA, **2005**, pp. 64-65.
[5] F. C. Chen, *Introduction to Plasma Physics*, Plenum, New York, **1977**.
[6] H. Yasuda, "*Plasma Polymerization*", Academic Press Inc. Ltd., London, **1985**, pp. 35-37.
[7] K. Samanta, M. Jassal, A. K. Agrawal, *Indian J. Fibre Text. Res.* **2006**, *31*, 83.
[8] S. Verprek, *Current Topics Mat. Sci.* **1980**, *4*, 151.
[9] D. B. Graves, *AIChE J.* **1989**, *35 (1)*, 1.
[10] R. H. Partridge, *J. Chem. Phys.* **1966**, *45*, 1685.
[11] N. Morosoff, B. Crist, M. Bumgarner, T. Hsu, H. Yasuda, *J. Macromol. Sci. Chem.* **1976**, *A10*, 451.
[12] J. Behnisch, J. Friedrich, H. Zimmermann, *Acta Polymerica* **1991**, *42*, 51.
[13] M. Hudis, L. E. Prescott, *Polym. Lett.* **1972**, *10*, 179.
[14] J. C. Ashley, *J. Electron Spectosc. Relat. Phenom.* **1988**, *46*, 199.
[15] G. Bonizzoni, E. Vassallo, *Vacuum* **2002,** *64*, 327.
[16] P. Krüger, R. Knes, J. Friedrich, *Surf. Coat. Technol.* **1999**, *112*, 240.
[17] D. P. Ames, S. J. Chelli, *Surf. Coat. Technol.* **2004**, *199-207*, 187.
[18] J. Janca, P. Stahel, J. Buchta, D. Subedi, F. N. Krcma, J. Pryckova, *Plasma. Polym.* **2001**, *6 (1/2)*, 15.
[19] M. M. Hossain, A. S. Herrmann, D. Hegemann, *Plasma Process. Polym.* **2007**, *4*, 135.
[20] M. Keller, A. Ritter, P. Reimann, V. Thommen, A. Fischer, D. Hegemann, *Surf. Coat. Technol.* **2005**, *200*, 1045.
[21] K. K. Wong, X. M. Tao, C. W. M. Yuen, K. W. Yeung, *Text. Res. J.* **2001**, *71 (1)*, 49.
[22] H. R. Yousefi, M. Ghoranneviss, A. R. Tehrani, S. Khamseh, *Surf. Interface Anal.*, **2003**, *35*, 1015.
[23] R. Molina, P. Jovancic, D. Jocic, E. Bertran, P. Erra, *Surf. Interface Anal.* **2002**, *35*, 128.
[24] S. Luo, W. J. van Ooij, *J. Adhes. Sci.* **2002**, *16 (13)*, 1715.
[25] G. Placinta, F. Arefi-Khonsari, M. Gheorghiu, J. Amouroux, G. Popa, *J. Appl. Polym. Sci.* **1997**, *66*, 1367.
[26] F. Arefi-Khonsari, M. Tatoulian, F. Bretagnol, O. Bouloussa, F. Rondelez, *Surf. Coat. Technol.* **2005**, *200*, 14.
[27] H. T. Oyama, J. P. Wightman, *Surf. Interface Anal.* **1998**, *26*, 39.
[28] W. J. van Ooij, S. Luo, S. Datta, *Plasma. Polym.* **1999**, *4 (1)*, 33.
[29] F. F. Chen, *Phys. Plasma.* **1995**, *2 (6)*, 2164.
[30] S. Morita, S. Hattori, *Pure Appl. Chem.* **1985**, *57 (9)*, 1277.
[31] S. Sigurdsson, R. Shishoo, *J. Appl. Polym. Sci.* **1997**, *66*, 1591.
[32] U. Vohrer, M. Müller, C. Oehr, *Surf. Coat. Technol.* **1998**, *98*, 1128.
[33] I. Tajima, M. Yamamoto, *J. Polym. Sci.: Polym. Chem.* **1985**, *23*, 615.
[34] M. G. McCord, Y. J. Hwang, Y. Qiu, L. K. Hughes, M. A. Bourham, *J. Appl. Polym. Sci.* **2003**, *88*, 2038.

[35] A. Bismarck, M. E. Kamru, J. Springer, J. Simitzis, *Appl. Surf. Sci.* **1999**, *143*, 45.
[36] D. Hegemann, *Indian J. Fibre Text. Res.* **2006**, *31*, 99.
[37] N. Inagaki, "*Plasma Surface Modification and Plasma Polymerization*", Technomic Publishing Co., Lanchester, PA, **1996**, pp. 127-130.
[38] H. E. Wagner, in: "*Low Temperature Plasma Physics*", eds. R. Hippler, S. Pfau, M. Schmidt, K. H. Schoenbach, Wiley-VCH, Berlin **2001**, pp. 305.
[39] A. Rutscher, H.-E. Wagner, *Plasma Sources Sci. Technol.* **1993**, *2*, 279.
[40] T. Wakida, S. Tokino, *Indian J. Fibre Text. Res.* **1996**, *21*, 69.
[41] F. Massines, A. Rabehi, P. Decomps, R. B. Gadri, *J. Appl. Phys.* **1998**, *83*, 2950.
[42] J. Pichal, J. Koller, L. Aubrecht, T. Vatuna, P. Spatenka, in *Plasma Polymers and Related Materials*, edited by M. Mutlu (Hacettepe University Press, Ankara, Turkey), **2005**, 201.
[43] H. Höcker, *Pure Appl. Chem.* **2002**, *74*, 423.
[44] J. Y. Kang, M. Sarmadi, *AATCC Rev.* **2004**, *11*, 29.
[45] S. L. Kaplan, *Proceedings, 41st Ann. Tech. Conf. SVC* (Boston, MA), **1998**, 345.
[46] C. I. Simionescu, F. Denes, M. M. Macoveanu, I. Negulescu, *Makromol. Chem.* **1984**, 8.
[47] D. J. Balazs, PhD thesis, École Polytechnique Fédérale De Lausanne, Thése N° 2748, **2003**.
[48] www.igb.fraunhofer.de
[49] H. J. Mathieu, *Surf. Interface Anal.*, **2001**, *32*, 3.
[50] R. J. Good, in *Contact Angle, Wettability and Adhesion* K. L. Mittal, Ed. VSP, Utrecht, Netherlands, **1993**, pp. 3-36.
[51] www.datacolor.com

3. Surface Wettability by Non-polymer Forming Plasmas

3.1. Introduction

The plasma treatment of fabrics has gained in importance over the past years as a textile finishing process for technical and medical textiles as well as for composite materials to improve their surface properties like wettability and adhesion.[1-8] Compared to conventional, wet-chemical textile finishing, plasma technology is also of increasing interest due to its environmental sustainability.[3,9]
Plasma processing, a developing field of applied physics and chemistry,[10] modifies the outermost layers of the fabric surface in the nanometer range, while the bulk properties as well as the quality of the material remain unaffected.[11-14] PET fabrics made of manmade-fiber reveal a moderate hydrophilic character due to the lack of polar groups. Moreover, hydrophobic sizes are generally used during the fabric production.[15] To obtain a good wettability, i.e. a high surface energy, some polar groups such as hydroxyl (-OH) or carboxyl (-COOH) should thus be formed at the fabric surface.[6,16] For the hydrophilization of fabrics, a low pressure and low temperature plasma treatment with oxygen-containing gases is expected to enable a good penetration of the textile structure due to long-living oxygen radicals.[17] Poll et al. considered the working pressure as the most important parameter in order to obtain a mean free path in the gas phase, which should be lower than textile distances.[11] Therefore, we performed RF plasma discharges with air, CO_2, water vapor, He/O_2 and Ar/O_2 mixtures by varying pressure, power and treatment time.
Contact angles (CAs) describing the liquid-vapor and liquid-solid interfaces are widely used for the study of the wetting/non-wetting phenomena on a solid material. Interfacial tension can be derived on flat, homogeneous surfaces by using liquids with different surface tensions.[18] For heterogeneous structures such as textile fabrics, however, the CA is not only affected by interfacial tension, but also by other phenomena such as surface roughness, chemical heterogeneity, polar groups, sorption layers, suction, porosity, swelling, molecular orientation, yarn tension etc. which complicate the direct measurement.[19-20] CA measurements on single fibers are described in numerous reviews and articles mainly based on the wetting force measurement by the Wilhelmy principle.[21-24] However, this process is not suitable for woven fabrics due to liquid uptake in the pore structure by capillary forces.[24]
In this study, the CA on a plasma-treated PET fabric was determined by observing the capillary rise as a function of time on vertically attached fabric strips with the lower end being immersed in the wetting liquid. According to Poiseuille's law the theoretical basis of this capillary rise test was described by Washburn.[25] The rate of liquid penetration into a porous body is given by:

$$\frac{dH}{dt} = \frac{R_D^2}{8\eta H}\left(\frac{2\gamma\cos\theta_A}{R_S} - \rho g H\right) \quad (3.1)$$

where H is the height reached by the liquid at time t, R_D the mean hydrodynamic radius of pores, η the viscosity of the liquid, γ and ρ the surface tension and density of the liquid on the solid, respectively, θ_A the advancing CA of the liquid on the solid, which is normally larger than the static one, and g the acceleration due to

gravity. R_S represents the mean static radius of pores which is a constant equal to the geometrical radius of pores. The hydrodynamic radius R_D, on the other hand, also depends on the tortuosity of the pores,[26] which is not directly related to geometrical considerations.[27]
In the early stages of the process, the hydrostatic pressure can be neglected, and the integration of Equation (3.1) yields the modified Lucas-Washburn Equation (3.2):

$$H^2 = \frac{R\gamma\cos\theta_A}{2\eta}t \tag{3.2}$$

where $R = \frac{R_D^2}{R_S}$ represents size of the capillary. It is also an important parameter which describes the capillary velocity within fabrics. Thus, a plot of H^2 vs. t should be linear giving the wicking coefficient or so-called capillary diffusion coefficient $D_C = \frac{R\gamma\cos\theta_A}{2\eta}$ that is related to the porosity of the fabric, and to the properties of the liquid ρ and η.[28]
At equilibrium, when capillary and hydrodynamic forces are equal, the maximum height is reached by the liquid at:

$$H_{Eq} = \frac{2\gamma\cos\theta_{Eq}}{R_S\rho g} \tag{3.3}$$

If the hydrostatic pressure can not be neglected ($H_{Eq} \cong \frac{2\gamma\cos\theta_{Eq}}{R_S\rho g}$, i.e. $\theta_A \cong \theta_{Eq}$) and using Equation (3.3), Equation (3.1) can be written as

$$\frac{R_D^2\rho g}{8\eta}dt = (\frac{H}{H_{Eq} - H})dH \tag{3.4a}$$

by substitution of $\cos\theta_{Eq}$ using Equation (3.3), which gives

$$\frac{R_D^2\rho g}{8\eta}t = H_{Eq}\cdot\ln\frac{H_{Eq}}{H_{Eq} - H} - H \tag{3.4b}$$

or more simply, $Ct = H_1$
R_D can be calculated from the slope of the straight line, where C is a coefficient that depends only on the size of the capillaries, and on the nature of the liquid.[29]
Considering that a liquid with a low surface tension such as hexane, decahydronaphthaline (decaline) or octane yields a complete wetting of the fabric (treated/untreated),[23,27] it can be assumed that the static CA θ_{Eq} equals zero, i.e. $\cos\theta_{Eq}$ is one, and

$$R_S = \frac{2\gamma}{H_{Eq}\rho g} \tag{3.5}$$

Thus, the mean static radius of pores can be determined through the capillary rise of a completely wetting liquid. Consequently, water CAs, both static (using Equation (3.3) and (3.5)) and advancing (using Equation (3.2), (3.4b) and (3.5)), on plasma-treated PET fabrics can be calculated to evaluate the degree of hydrophilicity using some modified forms of Lucas-Washburn's equation derived from Poiseuille's law, as discussed above.

At first, plasma process parameters were systematically optimized in this study by means of a suction test. Second, the capillary rise test was used to compare the apparent CAs within the textile fabric to PET foils treated by using the same plasma parameters. An aging study for both fabrics and foils was carried out with the aim of evaluating the penetration of the fabric structure by the plasma species. The chemical composition at the fabric surface was investigated by means of XPS.

3.2. Experimental

3.2.1. Plasma Treatment

The single layer, tightly woven poly(ethylene terephthalate) (PET) fabric (76 ends/inch, 76 picks/inch, 43.5 g/m^2) used in this study was supplied by Sefar Inc., Switzerland (Figure 1). The fabrics and foils (50 μm in thickness) were cut into pieces of 10x15 cm^2 for plasma processing. The untreated foils showed a static CA of $\approx$70°.

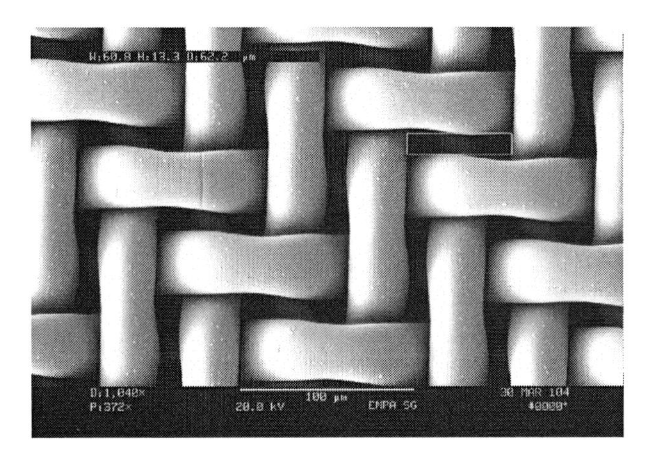

Figure 1 Scanning electron micrograph of the used PET fabric. The mesh opening is 13.3 μm wide.

In order to enhance the penetration of reactive plasma species into the textile structure in the micrometer range,[11] plasma treatments were carried out at low pressure to hydrophilize the fabric and foils. A capacitively coupled RF batch reactor (13.56 MHz) was used as schematically shown in Figure 2. A defined geometry and power coupling allowed reliable plasma treatments at varying conditions (power, pressure, treatment time, gases). Fabrics and foils were processed at the internal, driven electrode (10x15 cm^2), while the cylindrical recipient acted as grounded electrode. A shower head above the electrode enabled homogeneous treatments over the entire electrode area. The vacuum system composed of a rotary pump and a roots pump generated a base pressure of 10^{-2} Pa. The working pressure under gas flow was measured by a Baratron pressure gauge (MKS Instruments, Germany) which was adjusted by a throttling valve. V/I probe measurements (ENI model 1640) indicated an absorbed power of 70% and a ratio of (negative) bias V_{bias} to excitation voltage V_o of 0.8 due to the asymmetric set-up. In such discharges, the V_{bias} is added to RF excitation V_{rf}, Thus, V_{rf} (t) = V_o Cos (2πft) + V_{bias}, where (2πf) the angular frequency, and t the time.

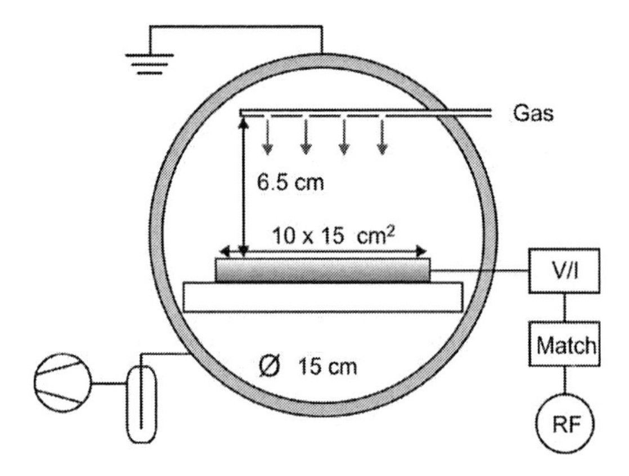

Figure 2. Schematic drawing of the plasma reactor used.

To optimize the plasma treatments an extensive parametric study was performed by varying important plasma parameters such as pressure (1.5-60 Pa), power (5-40 W), treatment time (0.5-5 min), and process gas (air, CO_2, water vapor, Ar/O_2 and He/O_2 as well as pure Ar). The plasma-treated samples were then kept in a conditioned room (20 ± 2°C, 65 ± 2% RH) for further experiments.

3.2.2. Suction Test

When the CA of water on a plasma-treated PET fabric is below a certain level, the water droplet is sucked in, which was the case after all plasma treatments performed. To investigate different plasma conditions, a simple way to measure wetted area was used in our laboratory and combined with a camera and a computer controlled program (MatLab). A normal syringe was used to drop a defined quantity of water (50 μL) onto the tensioned sample, and pictures of the spreading droplet were taken as a function of time to derive the wetted area.

3.2.3. Capillary Rise Test

For the capillary rise test a Camag chromatogram immersion device III (Germany) composed of a tank filled with the particular liquid, a micrometer scale and an adjustable carrier was used. The textile samples were cut into 15x2 cm^2 strips parallel to the warp direction. The strips were mounted in parallel to the scale, where its zero point maintained contact with the liquid surface in the tank, and were partly immersed into the liquid (at a depth of 1.5 cm). A light weight, which should not affect the geometrical structure of the sample, was placed at the end of the strip to keep it in a vertical position. The samples were conditioned before being tested. Bi-distilled water and decaline (Merck, Germany) were used for the capillary flow. The capillary rise was observed visually and the capillary heights were recorded every 5 seconds in the first minute, every 10 seconds in the following 6 minutes and after 1.5-2 hours the equilibrium height readings were also recorded both for decaline and water. It is noted that the decaline equilibrium heights were the same for every experiment, thus verifying the assumption of a complete wetting. Whenever the liquid front did not rise evenly, an average value was taken.

3.2.4. Material Characterization

The characteristics of the fabric surface were examined by SEM (Hitachi, S-4800) and ESCA (PHI LS 5600).

3.3. Result and Discussion

3.3.1. Plasma Treatment

In order to optimize the process parameters and to obtain a suitable gas or gaseous mixture, treatments were carried out at varying pressure, power, and exposure times. It is obvious that due to physical and chemical changes on fabric surfaces, plasma treatments have a significant effect on surface tension which contributes to the fabric wettability. Most of all, pressure has a strong influence on gas discharge, mainly its temperature, activation rate, flow rate, and plasma induced molecular fragmentation due to collisions resulting in excitation and recombination processes.[30] The variation in pressure at a fixed exposure time (1.5 min) and power (15 W) is shown in Figure 3.

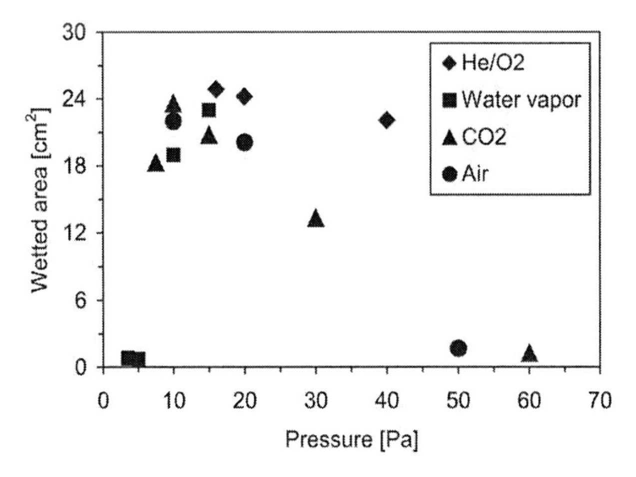

Figure 3. Effect of gas pressure on the suction of a water droplet into plasma-treated PET fabrics using different gases (1.5 min, 10 W).

The wettability of the fabric is mostly improved at low pressure (10-16 Pa) due to the more intense gas-phase molecular fragmentation. Conversely for a higher pressure, wettability is reduced significantly due to a low activation rate, which leads to a reduction in molecular fragmentation within the gas phase. It should be mentioned that oxygen containing gases were used, because long-living atomic oxygen radicals are able to penetrate into the textile fabric by surface diffusion, and the mean free path length of about 400 μm was long compared to the textile distances of the fabric (<15 μm).

The variation of power revealed that a power input of 10-15 W already yields a maximum wetted area, as can be seen from Figure 4. At a lower power of 5 W the formation of polar groups was slow at short exposure times resulting in low wettability.[13] At a fixed power (10-15 W) and pressure (10-16 Pa), as shown in Figure 5, the fabric wettability gradually increased with an increase in treatment time. However, a good hydrophilization effect was already obtained after a short plasma exposure. Moreover, from Figures 3-5 it was found that a He/O_2 mixture performed best compared to CO_2, air and water vapor with respect to the wetted area using the suction test. Ar/O_2 mixtures were found to be comparable to He/O_2 at the derived optimum parameters of 5 min treatment time, 16 Pa pressure, and 15 W power input for the hydrophilization of the used PET fabric. Addition of oxygen to Ar or He yields more excited species in the plasma zone such as long-living He* and Ar* metastables and long-living O atoms, which are able to penetrate into textile structures.

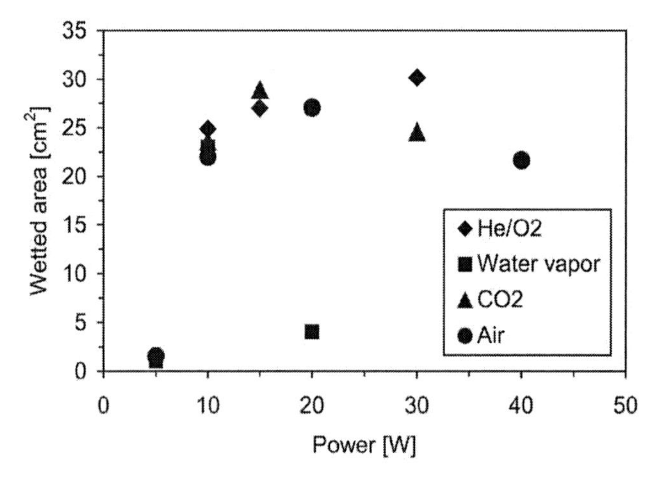

Figure 4. Effect of power input on the wetted area of plasma-treated PET fabrics using different gases (1.5 min, 10-16 Pa).

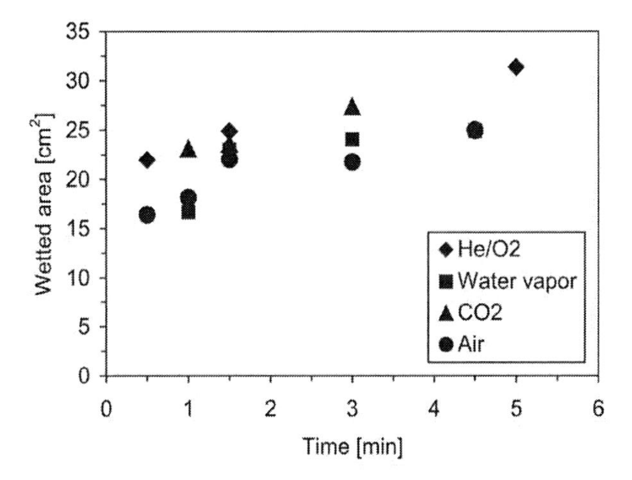

Figure 5. Water-wetted area of plasma-treated PET fabrics using different gases depending on treatment time (10 W, 10-16 Pa).

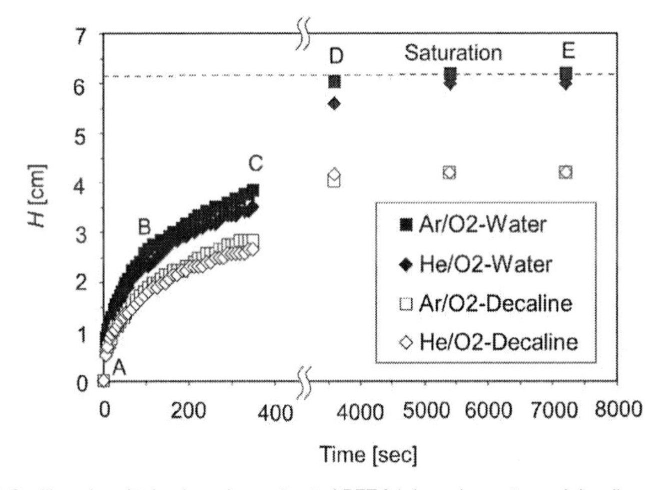

Figure 6. Capillary rise obtained on plasma-treated PET fabrics using water and decaline as liquid.

3.3.2. Capillary Rise Analysis

Capillary rise within a textile is influenced by a number of factors, especially fabric structure (yarn count, fabric density, weave design, porosity, fiber content etc.), which will be investigated in greater detail in a further study to be published elsewhere.[31] Here, the spontaneous liquid wicking of He/O_2 or Ar/O_2 plasma-treated PET fabrics is reported as a function of time. It is evident from the capillary rise H over time t as shown in Figure 6 that in all cases after plasma treatment the capillary rise with water is remarkably improved due to surface oxidation, whereas without treatment no rise can be obtained. In region A-B, the liquid wicks spontaneously into the fabric due the capillary pressure, where acceleration due to gravity in Lucas-Washburn's equation (Equation (3.2)) is neglected. The B-C region of the curves illustrates the subsequent capillary rise, which increases very slowly before reaching the equilibrium state (D-E). At point D the liquid front reaches its maximum height. When capillary and hydrodynamic forces are equal (D-E), no further rise is observed.

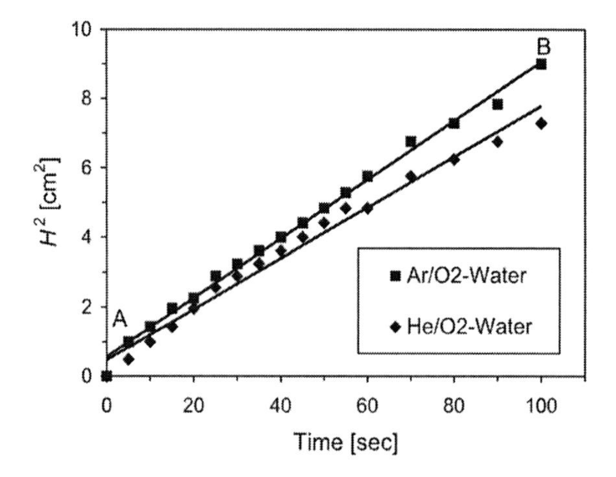

Figure 7. Square root of capillary rise on plasma-treated PET fabrics depending on time of rise.

It was found that the capillary rise and equilibrium height of decaline was the same for all experiments performed due to its excellent wetting properties. Thus, the cosine of θ_{Eq} (static CA) is one and the static radius of pores can be easily determined from Equation (3.5). In the spontaneous region of wicking (A-B), the derived curve from the capillary rise data is shown in Figure 7, showing that the square root of capillary rise H^2 is proportional to time t. This linear curve is valid for short times according to the Lucas-Washburn equation (Equation (3.2)). Under conditions of prolonged rising (region B-D) this equation becomes inadequate and hydrostatic pressure has to be taken into account. This behavior is well described by Zhmud et al.[32] Evidently, the analysis of the linear fit (H^2–t curve) is valid in the experimental time range of up to 100 seconds (A-B) for the examined fabric. If all data and equilibrium height are taken into account, straight lines are also obtained for H_1 vs. t (Figure 8) proving the validity of the presented evaluation, and the constant C can be determined from the slope of this curve as described by Equation (3.4b). Thus, hydrodynamic and static radii can be calculated yielding θ_A and θ_{Eq}.

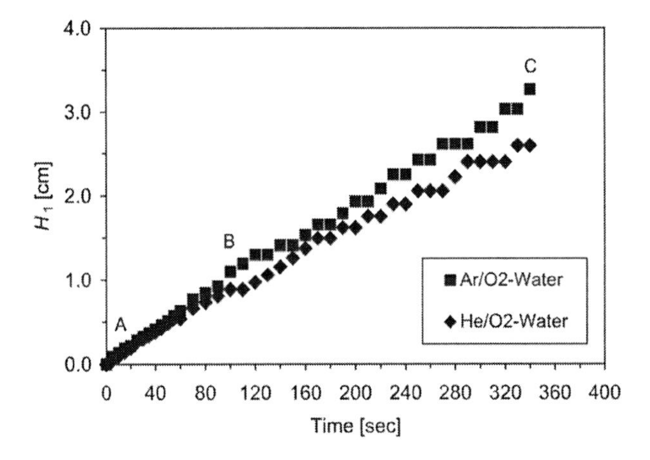

Figure 8. Linear dependence of H_1 values [see Equation (3.4b)] on time calculated from capillary rise data on plasma-treated PET fabrics.

3.3.3. Contact Angle

As described above, CAs, both static and advancing, can be obtained from the capillary rise test. It is clearly shown in Figure 9 that static CAs are smaller than the corresponding advancing ones (as it is expected). Both CAs increased gradually with storage time at ambient temperature indicating aging of the plasma-treated fabrics. The average static and advancing CA on fabrics obtained 2 hours after the plasma treatment were $\approx$44.0° and $\approx$57.5° respectively, whereas with a storage time of 8 days these values were attained within the range of 55-63° and 62-72°, respectively, where Ar/O_2 treatments exhibit a slightly better aging behavior. The static CA ratio between 2 hours and 8 days ageing is 1.3, which is similar to that found for the advancing one (1.2), i.e. the trend in changing CAs vs. aging is proportional. For longer periods (after 10 days) the capillary rise test can no longer be evaluated consistently due to a bad suction of water into the fabric at a water CA above 65°. Aging relies on reorganization processes on modified polymeric surfaces. For the improved performance of Ar/O_2 plasmas compared to He/O_2 plasmas a more crosslinked polymer surface might be assumed yielding a reduced reorientation of polar groups toward the textile bulk.[8]
For the comparison of the textile treatments absolute (static) CAs on foils were also measured after plasma treatment at the same conditions and the aging behavior was observed (Figure 10). Comparing the static CAs on fabrics and foils (Figures 9-10) it can be seen that the foils show a slightly better hydrophilization directly after plasma activation. However, the results indicate a good penetration into the textile structure using Ar/O_2 or He/O_2 plasmas. Especially, the aging behavior was found to be very similar.

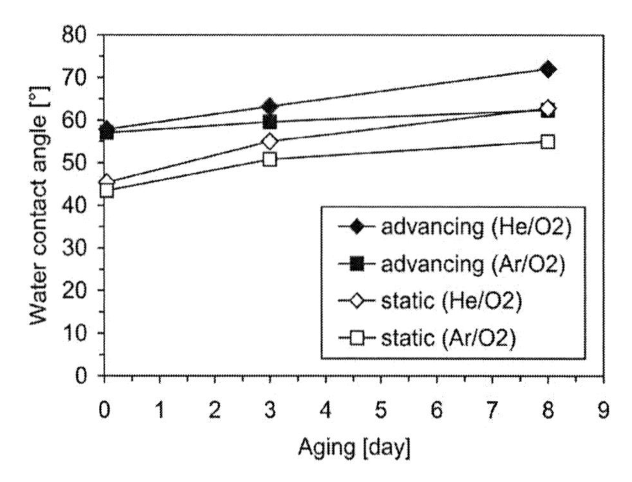

Figure 9. Increase of water contact angle (static and advancing as obtained from capillary rise test) with aging for different plasma-treated PET fabrics (5 min, 16 Pa, 15 W).

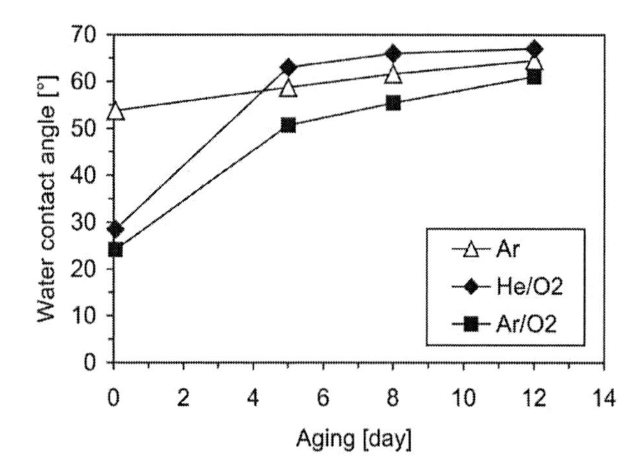

Figure 10. Increase of water contact angle (static) with aging for different plasma-treated PET foils (5 min, 16 Pa, 15 W).

A pure Ar plasma, on the other hand, which results in a noticeable hydrophilization of PET foils (Figure 10), yields no pronounced treatment of the PET fabric, i.e. a capillary rise can not be detected. Hence, it can be inferred that oxygen radicals play an important role in the penetration of the textile structures at the pressure range used (10-16 Pa), whereas reactive Ar particles (ions and neutrals) show a too long mean free path and a too low reactivity to effectively activate the covered fiber areas of the PET fabric hindering the capillary rise.

3.3.4. Surface Characterization

While SEM pictures show no remarkable change of the rather smooth fiber surfaces after plasma treatment, ESCA reveals an increase in oxygen-containing functionalities and a cleaning of the fiber surfaces (Figure 11) which is responsible for the obtained hydrophilization by carboxyl, carbonyl and hydroxyl groups. Compared to a reference spectrum of PET, the fabric surface reveals a nanometer thick layer (regarding the information depth of ESCA of ≈5 nm) of hydrocarbons, probably mineral oils, used as sizing during fabric production.

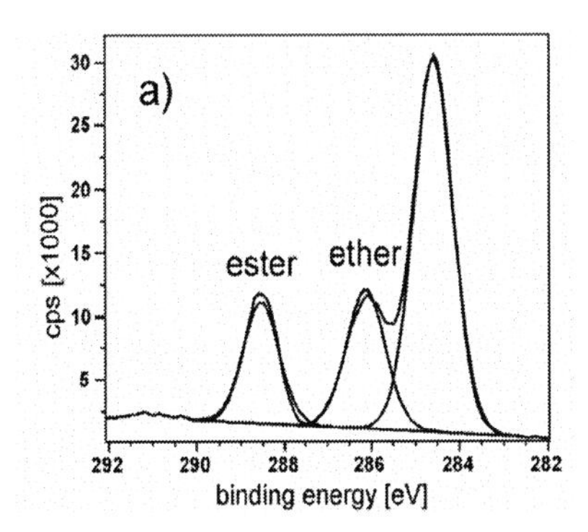

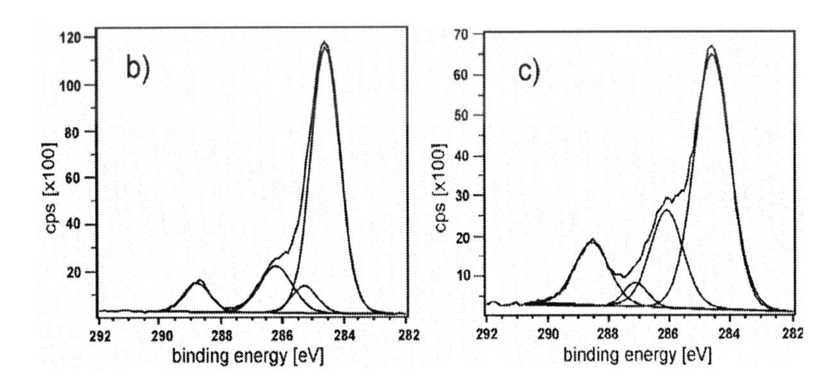

Figure 11. ESCA C1s spectrum of a reference PET sample (a), the untreated PET fabric (b), and the plasma-treated PET fabric (c). Ar/O_2 plasma at 5 min, 16 Pa, and 15 W was used.

3.4. Conclusion

The hydrophilization of a PET textile fabric was investigated using low pressure RF plasmas. Depending on the process gas used, the wettability and capillarity of PET fabrics can be significantly improved, while the results are strongly correlated with

the plasma parameters (especially pressure and power) which were optimized using a suction test. Various plasma treatments with air, CO_2, water vapor, He/O_2 and Ar/O_2 as well as different treatment conditions were examined. The evaluation of a capillary rise test by using modified forms of Lucas-Washburn's equation enables the obtaining of both static and advancing CAs within the textile structure. Thus, the treatment on fabrics can be compared to the treatment on flat substrates like foils and the penetration of reactive plasma species into textile structures can be investigated. It was found that oxygen containing plasmas (mixtures of Ar or He with O_2) revealed a similar wetting and aging behavior achieved within the PET fabric compared to a PET foil under the same processing conditions (5 min, 16 Pa, 15 W). A pure Ar plasma, on the other hand, yielded a worse hydrophilization of the fabric indicating the importance of oxygen radicals for the penetration into tightly woven fabrics. Reactive Ar species have a more vertical directed interaction on the fabric surface where the inter-yarn surface and overlapping zone (the place where the warp and weft yarn are in close contact) remain untreated and thus decay wettability and capillarity. Therefore, the gas selection took priority over other parameters regarding the plasma penetration into the heterogeneous textile structure.
However, the modified polar surface is not stable, since aging due to reorganization processes was obtained. Nevertheless, capillary rise and suction of the plasma-treated PET fabric could be observed over a period of at least eight days, which is of importance for subsequent processing steps such as coating, lamination, coloration or other wet-chemical processes.[33,34] Hence, plasma activation using non-polymerizable gases is appropriate for cleaning and activation of textiles in order to improve subsequent process steps such as lamination or plasma coating. To obtain more permanent hydrophilic treatments on textiles, the deposition of nanoscaled functional plasma coatings is required, which we have demonstrated recently.[35-36] Furthermore, the degree of plasma activation and plasma hydrophilization are closely linked to the textile structure and weave construction, which elaborately described in the next chapter.

3.5. References

[1] M. M. Hossain, D. Hegemann, P. Chabrecek, A. S. Herrmann, *J. Appl. Polym. Sci.* **2006**, *102*, 1452.
[2] D. Bechter, St. Berndt, R. Greger, W. Oppermann, *Technische Textilien* **1999**, *42*, 14.
[3] T. Okuno, T. Yasuda, H. Yasuda, *Text. Res. J.* **1992**, *62 (8)*, 474.
[4] S. Ruppert, B. Müller, T. Bahners, E. Schollmeyer, *Textil Praxis Int.* **1994**, *48*, 614.
[5] W. G. Pitt, J. E. Lakenan, A. B. Strong, *J. Appl. Polym. Sci.* **1993**, *48*, 845.
[6] T. Wakida, S. Tokino, *Indian J. Fibre Text.* **1996**, *21*, 69.
[7] S. Carlotti, A. Mas, *J. Appl. Polym. Sci.* **1998**, *69*, 2321.
[8] B. Z. Jang, *Compos. Sci. Technol.* **1992**, *44*, 333.
[9] H. U. Poll, S. Schreiter, *Melliand Textilberichte* **1998**, *6*, 466.
[10] A. Raffaele-Addamo, C. Riccardi, E. Selli, R. Barni, M. Piselli, G. Poletti, F. Orsini, B. Marcandalli, M. R. Massafra, L. Meda, *Surf. Coat. Technol.* **2003**, *174-175*, 886.
[11] H. U. Poll, U. Schladitz, S. Schreiter, *Surf. Coat. Technol.* **2001**, *142-144*, 489.
[12] F. Ferrero, *Polym. Test.* **2003**, *22*, 571.
[13] B. Gupta, J. Hilborn, CH. Hollenstein, C. J. G. Plummer, R. Houriet, N. J Xanthopoulos, *Appl. Polym. Sci.* **2000**, *78*, 1083.
[14] D. Hegemann, H. Brunner, C. Oehr, *Nucl. Instrum. Meth. B* **2003**, *208*, 281.
[15] M. Keller, A. Ritter, P. Reimann, V. Thommen, A. Fischer, D. Hegemann, *Surf. Coat. Technol.* **2005**, *200*, 1045.
[16] C. Jie-Rong, W. Xue-yan, W. Tomiji, *J. Appl. Polym. Sci.* **1999**, *72*, 1327.
[17] C. Oehr, D. Hegemann, M. Müller, U. Vohrer, M. Storr, in: Plasma Processes and Polymers, eds. R. d'Agostino, P. Favia, C. Oehr, M.R. Wertheimer, Wiley-VCH, Weinheim, **2005**, pp. 309.
[18] M. L. Steen, A. C. Jordan, E. R. Fisher, *J. Membrane Sci.* **2002**, *204*, 341.
[19] M. Maejima, *Text. Res. J.* **1983**, *53*, 427.
[20] A. Perwuelz, M. Casetta, C. Caze, *Polym. Test.* **2001**, *20*, 553.
[21] D. Saihi, A. El-Achari, A. Ghenaim, C. Caze, *Polym. Test.* **2002**, *21*, 615.
[22] Y. Hsieh, B. Yu, *Text. Res. J.* **1992**, *62 (11)*, 677.
[23] Z. Persin, K. Stana-Kleinschek, T. Kreze, *Croatica Chemica Acta* **2002**, *75 (1)*, 271.
[24] C. Rulison, Krüss Laboratory Services and Instrumentation for Surface Science, Technical Note #302.
[25] E. W. Washburn, *Physical Review* **1921**, *17 (3)*, 273.
[26] A. Siebold, M. Nardin, J. Schultz, A. Walliser, M. Oppliger, *Colloid. Surf. A* **2000**, *161*, 81.
[27] A. Siebold, A. Walliser, M. Nardin, J. Schultz, *ICSI-CNRS*, France and MBT Ldt, Switzerland, 645.
[28] Y. K. Kamath, S. B. Hornby, H. -D. Weigmann, M. F. Wilde, *Text. Res. J.* **1994**, *64 (1)*, 33.
[29] A. Perwuelz, P. Mondon, C. Caze, *Text. Res. J.* **2000**, *70 (4)*, 333.
[30] A. M. Sarmadi, T. H. Ying, F. Denes, *Text. Res. J.* **1993**, *63 (12)*, 697.
[31] M. M. Hossain, A. S. Herrmann, D. Hegemann, *Plasma Process. Polym.* **2006**, *3*, 299.
[32] B. V. Zhmud, F. Tiberg, K. Hallstensson, *J. Colloid. Interf. Sci.* **2000**, *228*, 263.
[33] D. Hegemann, C. Oehr, A. Fischer, *J. Vac. Sci. Technol. A* **2005**, *23 (1)*, 5.

[34] D. Hegemann, *Adv. Eng. Mater.* **2005**, *7 (5)*, 401.
[35] D. Hegemann, *Indian J. Fibre Text.* **2006**, *31*, 99.
[36] M. M. Hossain, D. Hegemann, A. S. Herrmann, *Plasma Process. Polym.* **2007**, *4*, 471.

4. Plasma Species Penetration into Textile Structures

4.1. Introduction

Plasma techniques have gained much importance for technological applications for several reasons: it generates radicals and oxidizes the surface (i.e. hydrophilic surface);[1-3] it changes topography: adhesion, repellence properties, roughening of the surface;[4-13] it permits the cleaning of surfaces which leads to an increase of quality printing, dye-uptake, painting etc.[14-16]
Textile structure and construction are in turn dependent on many factors such as the type of weave pattern, weight (g/m^2) of fabric, type of the fiber content, fiber fineness, ends/inch, picks/inch, and also yarn parameters like the twist factors. In comparison with flat surfaces (foils, polymer solids etc.) fabrics have complex architectures which in fact consist of two surfaces, one which is macroscopic and visible by the naked eye and another which is the actual inner surface comprised of the inter-fiber/filament space, the inter-yarn space and the pore size distribution. Thus, because of such heterogeneous complex structure, which makes the plasma activation more complicated than for solid polymeric materials, the plasma treatments on textile fabrics are limited.[17,18] Nevertheless, several plasma processes are already commercialized within the textile industry.[19]
While abundant research articles are published on the effects of low pressure plasma on textile fabrics, there have been very few studies on the plasma species penetration into fabric structures. Therefore, in this work I have extended the studies to include the influence of plasma activation on the physico-chemical changes occurring at the textile surface of different polyester (PES) textile structures (woven/knitted) by using oxygen containing gaseous mixtures (Ar/O_2 and He/O_2) as shown in Chapter 3. Recently, it is proved that these gaseous mixtures are most effective to clean and activate the PES textiles.[20] Therefore, capillary rise tests has been carried out in order to determine the contact angle (CA) on fabrics, and thus the wetting properties of the modified fabric surfaces. In this manner, the correlation between the surface effects induced by plasma and the degree of plasma species (fragmented/charged gas particles) penetration into textiles can be understood. Aging studies have also been performed in order to examine the durability of the modified surfaces.
The most practical way to study wetting phenomena is the measurement of CAs of well defined liquids on a solid surface. However, serious problems arise in case of porous materials, which are of considerable technological interest. In the previous chapter, an alternative approach was demonstrated, based on the modified Washburn equation, which can be used to measure CAs by capillary rise. The interaction of liquids with fibrous assemblies is normally described by wicking, liquid transport. Liquid-wicking into textile fabrics is further complicated by the surface roughness, the heterogeneity, the diffusion of liquid into the fiber, and the capillary action of the fiber assemblies.[21,22] The geometrical features of fabrics such as thickness and density of fabric, twist, yarn types, and count etc. as well as internal volume and pore size distribution make the wetting phenomena of fabrics non-ideal.[23] Wetting is a prerequisite for wicking; if a liquid does not wet, the liquid can not wick into a fabric with capillary spaces. By capillary forces a liquid can wick into

the inter-filament spaces spontaneously in a fabric that is made from continuous filament yarns, better than for twisted staple yarns.[24]

4.2. Experimental

4.2.1. Materials

Three different types of PES fabrics were investigated in this study, one of them is a rib knitted fabric and the other two are woven fabrics. The fabrics used in this work were supplied by AATCC, USA (adjacent PES), Nya Nordiska Textiles GmbH, Germany (sassa PES) and Empa Test Materials, Switzerland (knit PES). The fabric specifications are stated in Table 1. SEM images of different textile patterns were obtained using Hitachi, S-4800 with an accelerating voltage of 20 kV.

Table 1. Fabric specifications.

Name	Fabric structure	Weave design	Fiber content	Yarn type	Ends/in or Wales/in	Picks/in or Course/in	Weight (g m^{-2})
Adjacent PES	Woven	Plain	100% PES	Staple, twisted	46	49	130±5
Knit PES	Knit	Rib knit	100% PES	Staple, twisted	48	90	147±5
Sassa PES	Woven	Plain	100% PES	Filament, Non-twisted	77	157	64±5

4.2.2. Modification of Fabrics

Plasma treatments were carried out under low pressure in a batch plasma reactor which is shown in Figure 1. The set-up of the reactor is also described in the previous chapter.[25]

The sample was placed horizontally (no creases/folds) for the maximum contact with the electrode in the centre of the reactor. Before introducing process gases in the chamber, it was first evacuated to less than 0.1 Pa. To improve the hydrophilization effect, oxygen containing gas mixtures He/O_2 and Ar/O_2 (purity 99.99 vol.-%) were used in this investigation. The gaseous mixtures were introduced into the chamber at flow rates of 50 sccm Ar or He and 5 sccm O_2. The working pressure was adjusted to 16 Pa, and then the plasma was generated at 15 W RF power for 5 min. The plasma-treated samples were then kept in a conditioned room (20 ± 2°C, 65 ± 2% RH) for further experiments.

Figure 1. Plasma batch reactor used.

4.2.3. Capillary Rise Test

A Camag chromatogram immersion device III (Germany) was used for the capillary rise tests. The system and experimental details are elaborately described in Chapter 3.

4.3. Result and Discussion

4.3.1. Modification and Plasma Species Penetration

The modification of a textile surface mainly depends on the gas nature, the pressure, and the exposure time. The experimental data found in the literature[26] showed that long-living oxygen radicals mixed with inert gas (He and Ar) lead to an increase of hydrophilic effect while a small amount of oxygen (He/Ar:O_2=10:1) resulted in a strong hydrophilic effect and more crosslinked surface in textiles.[3,27] The optimization of the process parameters were performed as described in Chapter 3.[28] It is suggested that the building-up of the polar groups was higher at moderate power (15 W) showing a higher activation rate, while polymer damage (degradation) can be avoided. The hydrophilicity was found to increase gradually with treatment time for polyester textiles. However, while a good hydrophilization effect can already be obtained after a short time (1 min),[13] in this study experiments were carried out with 5 min exposure time to enhance the plasma species penetration.

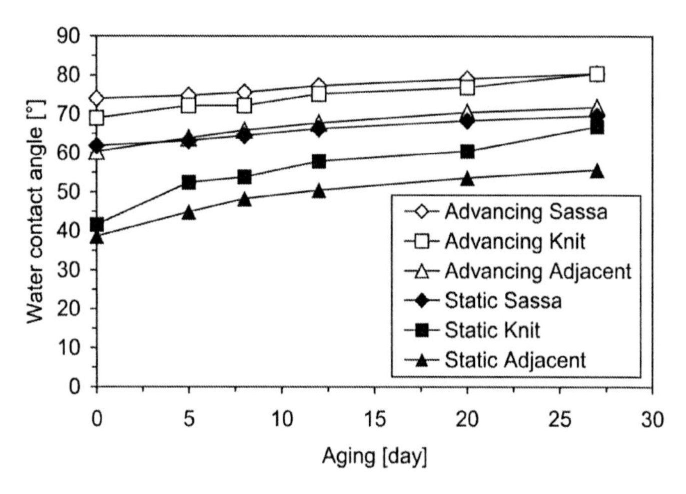

Figure 2. Water contact angles (static and advancing as obtained from capillary rise test) on Ar/O_2-plasma-treated PES fabrics as a function of aging time. Note: cleaned PES fabrics show a static contact angle of ≈75°-80°.

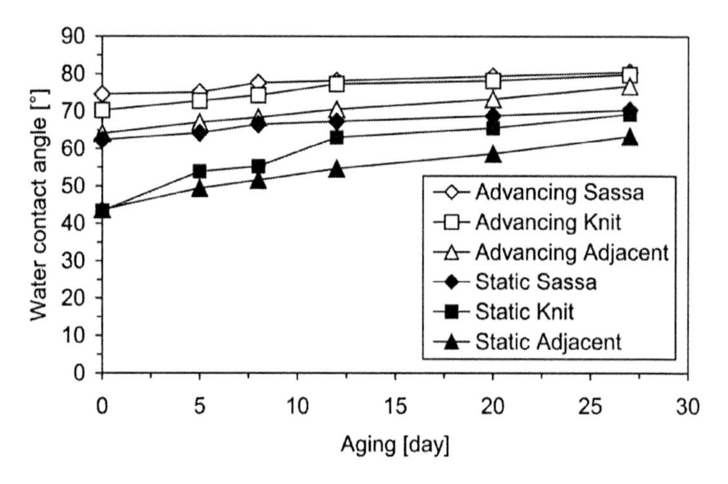

Figure 3. Water contact angles (static and advancing as obtained from capillary rise test) on He/O_2-plasma-treated PES fabrics as a function of the aging time. Note: cleaned PES fabrics show a static contact angle of ≈75°-80°.

Addition of small amounts of oxygen to Ar or He in the feed gas generates more excited species in the plasma zone such as long-living He* and Ar* metastables, and long-living O atoms, which are able to penetrate into textile structures. In general, no major changes are expected during the modification-reactions for both plasmas. Small differences might be found due to the energetic particle bombardment, the electron temperature and the density as well as the VUV radiation. In Figure 2 and 3 no significant difference between Ar/O_2 and He/O_2 plasmas in respect of CA can be seen. It is evident from these data that low pressure plasma has proven to be very efficient for textile fabric treatments,[3,17] whether the structure is knitted or woven. The low pressure plasma assures a more intense high electron energy tail and consequently, a more efficient gas phase molecular fragmentation.[18] The penetration of these fragmented gas particles into the voluminous textile is also influenced by the textile geometry, such as the distance between filaments/fibers in yarn and by the distance between two yarns (warp and filling yarns) in a fabric. Poll et al.[4] described these typical distances for single filaments from 1 to 10 µm and for yarns from 0.1 to 1 mm. At low pressure, the mean free path (the distance traveled by a molecule or atom between successive collisions) in the gas phase is higher than textile distances, and the collision of gas molecules with the fiber surface is enhanced as compared to gas-gas collisions, thus favoring a good penetration of plasma species into the textiles, especially when surface diffusion is predominant (low sticking coefficients for incoming reactive particles). In the case of higher pressure, the opposite phenomenon is observed, and the losses due to gas-gas collisions increase giving the gas radicals less opportunity to reach the reaction sites of the fiber/filament.

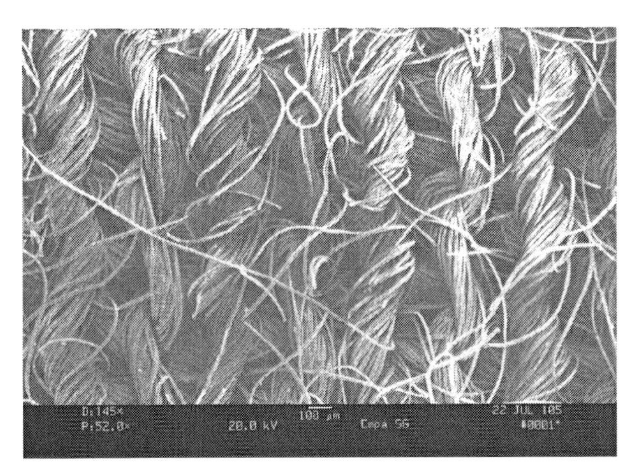

Figure 4. SEM micrograph of the knitted PES fabric.

In addition to the degree of plasma species penetration, the wettability is also attributed to the weave pattern. Specifically, the obtained hydrophilization of the looser structured fabrics, such as plain weave adjacent fabric and rib knitted fabric, has improved significantly. Due to reorganization processes at the surface, aging effects occurred yielding hydrophobic recovery (Figure 2 and 3). In particular, aging of the densely woven sassa fabric led to CAs comparable to the untreated PES ($\approx$75-80°), whereas the looser structured fabrics maintain their hydrophilic nature even after several weeks. In looser structured fabrics, energetic particles are assured to move into the inter-yarn spaces, inter-fiber spaces, and overlapping areas of warp and filling yarns more easily, that is to say that the plasma species bombardment taking place with a higher surface area caused enhanced penetration into the textiles. This resulted in the desired surface modification and surface activation. Conversely, in the case of the tightly woven sassa fabric, a rather poor hydrophilization effect was found, since the plasma species penetration into the textile structure was limited and the fabric shows a lower surface area (see Table 1).

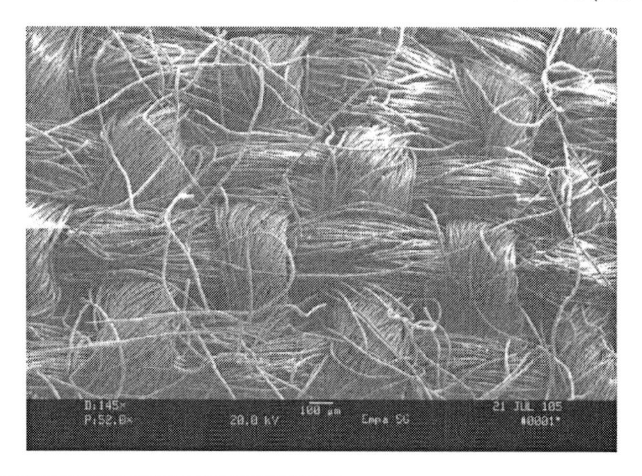

Figure 5. SEM micrograph of the adjacent PES fabric.

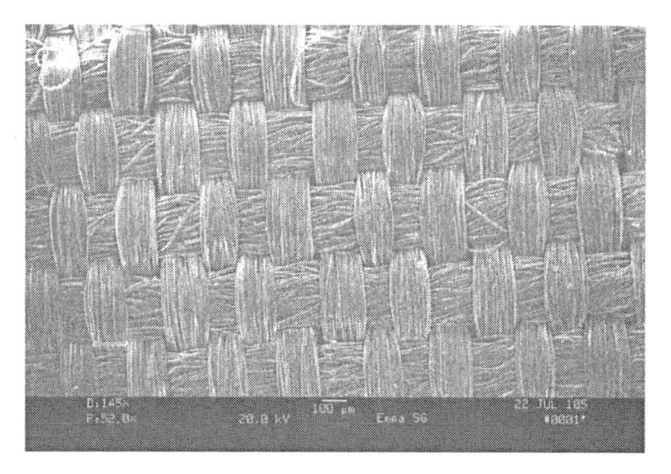

Figure 6. SEM micrograph of the sassa PES fabric.

4.3.2. Wettability and Contact Angle of Fabrics

Wetting of fibrous assemblies, such as fabrics, is a complex process. The surface wettability is directly related to the surface functional groups or to the surface energy, the surface roughness and the architecture of fabrics.[29] During low pressure plasma treatments, the improved wettability is attributed to the increase in the polar groups, and the surface oxidation, in particular in the case for case of textiles.[30,31] In general, polyesters show a strong hydrophobic behavior, and a low surface energy.

In all cases, the wettability of plasma-treated PES fabrics was improved noticeably due to the formation of several types of hydrophilic groups (for example -OH, -OOH, -COOH etc.)[3,28,32,33] on the fabric surface during plasma, or through post-plasma reactions, whereas the wettability was very poor for untreated samples.
CA measurements are a useful tool for surface characterization because they are determined by both wetting and capillarity properties. The Figure 2 and 3 showed that for the looser structured patterns (knit and adjacent fabrics) the CA was lower than that of tightly woven sassa fabric. This fact has suggested that the energetic particles and the long-living radicals impact mostly on the accessible fiber surfaces of the textiles resulting in higher wettability and higher capillary velocity. Moreover, a higher amount of projecting fibers (see Figure 4 and 5) on the adjacent and knitted fabrics facilitates higher capillarity and wicking. In the case of sassa fabric, relatively higher CA we obtained can be attributed to a lower amount of hydrophilic groups formed at the total fabric surface. The capillary flow between filaments and yarns in sassa fabric was found to be low. It can be concluded that the formation of polar groups on the textile surfaces strongly depends on textile architecture, while using the same fiber content (polyester).

4.3.3. Capillary Size and Capillarity

The capillary size and the capillarity is calculated according to Equation (3.2) to (3.5) (see Chapter 3). The capillary size is influenced by the fabric structure such as the filament/fiber distances, the inter-yarn spaces, the mesh/loop openings, and the roughness, as well as by the surface hydrophilicity. In all cases, the mean static radius of pores, which merely depends on textile geometry, was determined based on the total wetting liquid in order to obtain the capillary size and the water CA on fabrics. The geometry of the different textiles can be seen from Figure 4 to 6. An increase in the surface hydrophilicity is generally accompanied by an increase in the capillary size.[34]
Figure 7 illustrates that the capillary size of adjacent fabric was found to be higher as compared to the other two types of fabric. The sassa fabric has the smallest capillary size for both Ar/O_2 and He/O_2 plasma treatments while is also correlated to the highest CAs obtained. The capillary size of adjacent fabric for both plasma treatments was almost identical, as expected, and the same result has been found for sassa fabric. On the contrary, we observed noticeable differences with the knitted fabric. Moreover, aging effects were investigated for all three textiles over 4 weeks as shown in Figure 8 and 9. While the capillary size was almost constant for the woven fabrics (adjacent and sassa), the situation was different for the case of the knitted fabric. Since the change in CA with the aging was rather low, the (apparent) average capillary sizes in knitted fabric between 0 and 27 d of storage were found to be remarkably reduced as compared to those of the woven fabrics due to a higher irregularity in the capillary spaces (Figure 4). This result demonstrates that the CA determination using Lucas-Washburn's law is not ideal for knitted fabric due to its structural complexity yielding inconsistent aging effects. In contrast, the capillary size in sassa and adjacent fabrics was almost the same during aging for 27 d indicating a homogeneous and spontaneous wicking proving the validity of our approach to determine CAs on textile fabrics.

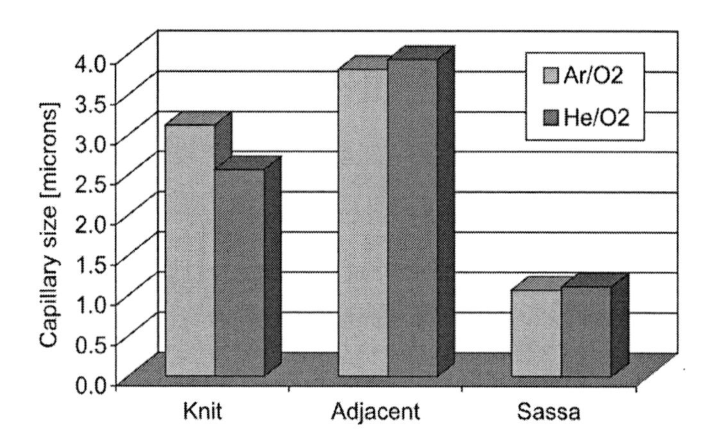

Figure 7. The evolution of the average capillary size (obtained by Lucas-Washburn's equation) of the plasma-treated (Ar/O_2 and He/O_2) PES fabrics during 27 d of aging.

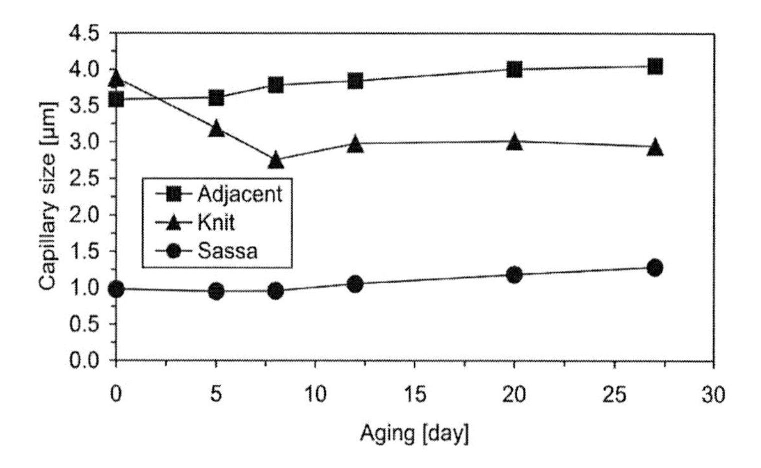

Figure 8. The evolution of the capillary size of the Ar/O_2-treated fabrics with the aging time.

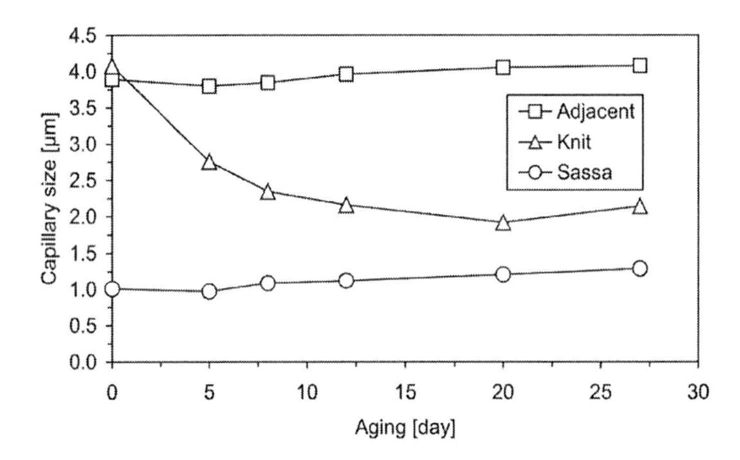

Figure 9. The evolution of the capillary size of the He/O_2-treated fabrics with the aging time.

The movement of a liquid in a fibrous assembly, especially in knitted fabrics, is discontinuous because of the inhomogeneous capillary system, and, in a consequence, the regression line does not pass through the origin (non-linear fit in H^2 versus t curve in Figure 10). Furthermore, the capillary flow of the liquid into the knitted fabrics is more complicated because of its loops, whereas woven fabrics consist of straight yarns, where fibers are more or less organized in a parallel array. Due to spontaneous wicking and capillarity, the square root of capillary rise (H^2) increases linearly with increasing the time (t) (linear fit in H^2 versus t in Figure 11). A knitted fabric structure is constructed by interlooping one or more ends of yarn, whereas woven fabrics are made of interlacing of warp and filling yarns. Theoretically, the capillary rise in a textile is produced by the organization of the parallel capillaries, which is in fact a capillary system composed of individual fibers/filaments in yarns. The woven fabrics consist of straight yarns containing fibers/continuous filaments laid more parallel to each other yielding a higher axial liquid flow in the vertical direction (warp direction) and transverse flow in the horizontal direction (weft direction). The liquid flow is slightly interrupted in the interlacing zone only (Figure 12) that might have some influences on the CAs. On the other hand, the flow front within the knitted structured fabric was either inhomogeneous, or the liquid wicking did not occur spontaneously (due to absent of fully axial or transverse yarns) resulting in a complex phenomena of capillarity and wicking. Thus, the capillarity in knitted fabric was irregular as compared to woven fabrics, although the knitted fabric could be plasma-activated due to its more open spaces.

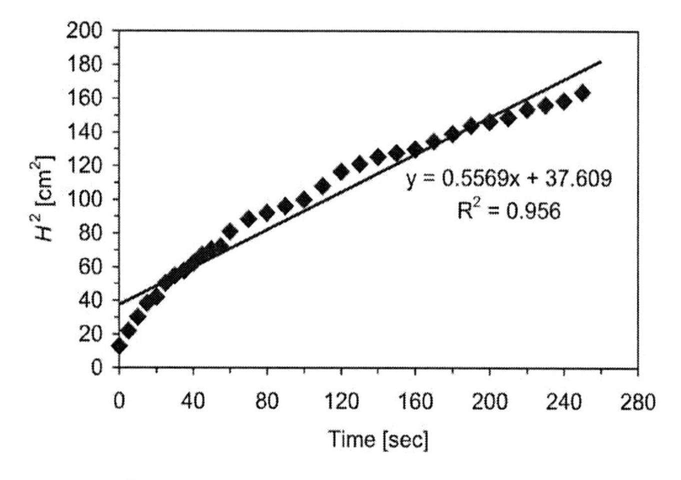

Figure 10. Non-linear H^2 versus t curve for the case of water on the plasma-treated knitted fabric (5 d aging) obtained by using Lucas-Washburn's equation.

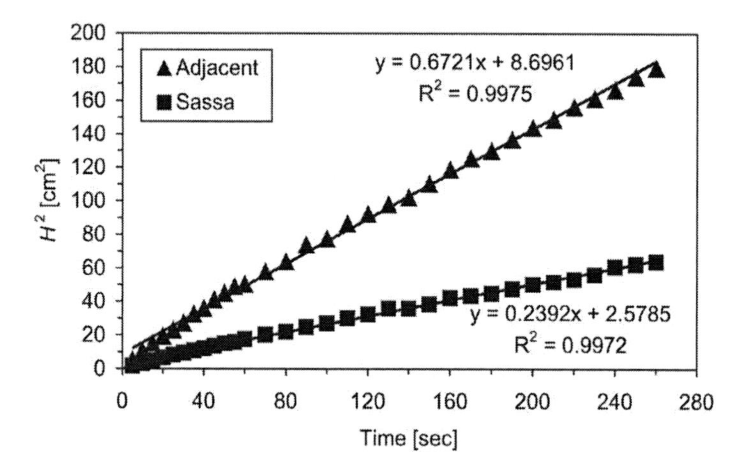

Figure 11. Linear fit of H^2 versus t curve for the case of water on the Ar/O_2-treated woven fabrics (5 d aging) obtained by using Lucas-Washburn's equation.

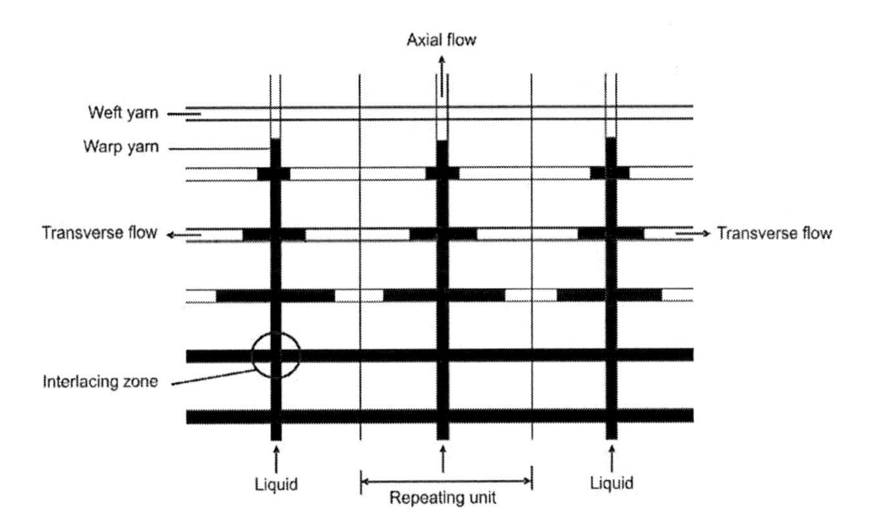

Figure 12. Simplified representation of the liquid flow in the plain weaved fabrics.

4.4. Conclusion

The hydrophilization effect on different fabric structures was investigated using low pressure RF plasmas. By using capillary rise data and the modified forms of Lucas-Washburn's equation, the determination of static and advancing CAs on fabrics enabled the characterization of the wettability, and of the degree of hydrophilicity of textiles. It was found that the loops in knitted fabric yielded some complications in determining the capillary rise. More specifically, the flow of the liquid front was interrupted in the vertical, as well as the horizontal direction of the fabric as a result of the inhomogeneous wicking. Thus, the use of Lucas-Washburn's law is not ideal for knitted fabrics. In the case of woven fabrics, axial (lengthwise direction) and transverse flow (width wise direction) occurred, thus the capillarity was homogenous favoring the spontaneous wicking and the transport of the liquid. Therefore, the hydrophilicity (wettability, wicking and capillarity) of woven fabrics can easily be characterized by the CAs obtained from the capillary rise test.

It is evident that low pressure plasma treatment is an effective way for surface modification of textiles. Both, Ar/O_2 and He/O_2 plasmas were found to yield good hydrophilization of textiles showing different structures. At low pressure (15 Pa), the higher mean free path assured minimum losses through gas-gas collisions, while the interaction of reactive gas particles with the reaction sites of the fibers was enhanced due to the low gas density. Therefore, the penetration of plasma species into the textile structure is enabled. In addition, the penetration of long-living radicals and of energetic particles is also significantly influenced by the textile geometries. The wettability of looser structured weaved patterns, such as the adjacent fabric and the knitted fabric, was improved markedly. Moreover, in the looser structured fabrics, the reactive gas particles easily moved into inter-yarn and –fiber spaces in fabrics resulting in improved wettability or capillarity. However, plasma-activated fabric surfaces were not durable over time due to reorganization processes on the surface. Conversely, in tightly woven sassa fabric the plasma species' impacts that occurred with a reduced surface area (and higher covered areas) led to an insufficient surface modification. These findings suggested that both the weave construction and the yarn type are two important parameters, in addition to pressure, power and time, which must be considered when performing surface activation of textiles, as the resulting wettability and hydrophilicity can vary quite significantly. Furthermore, plasma species penetration also plays a major role in plasma deposition, i.e. plasma polymerization, on textile fabrics as shown in the following chapters.

4.5. References

[1] M. M. Hossain, A. S. Herrmann, D. Hegemann, *Plasma Process. Polym.* **2006**, *3*, 299.

[2] S. Ruppert, B. Müller, T. Bahners, E. Schollmeyer, *Textil Praxis Int.* **1994**, *9*, 614.

[3] Y. J. Hwang, J. S. An, M. G. McCord, S. W. Park, B. C Kang, *Fibre Polym.* **2003**, *4 (4)*, 145.

[4] H. U. Poll, U. Schladitz, S. Schreiter, *Surf. Coat. Technol.* **2001**, *142-144*, 489.

[5] B. Gupta, J. Hilborn, CH. Hollenstein, C. J. G. Plummer, R. Houriet, N. Xanthopoulos, *J. Appl. Polym. Sci.* **2000**, *78*, 1083.

[6] D. Hegemann, H. Brunner, C. Oehr, *Nucl. Instrum. Meth. B* **2003**, *208 B*, 281.

[7] W. G. Pitt, J. E. Lakenan, A. B. Strong, *J. Appl. Polym. Sci.* **1993**, *48*, 845.

[8] T. Wakida, S. Tokino, *Indian J. Fibre Text.* **1996**, *21*, 69.

[9] H. Krump, M. Simor, I. Hudec, M. Jasso, A. S. Luyt, *Appl. Surf. Sci.* **2005**, *240*, 268.

[10] S. Carlotti, A. Mas, *J. Appl. Polym. Sci.* **1998**, *69*, 2321.

[11] B. Z. Jang, *Compos. Sci. Technol.* **1992**, *44*, 333.

[12] D. Hegemann, C. Oehr, A. Fischer, *J. Vac. Sci. Technol.* **2005**, *23 A*, 5.

[13] D. Hegemann, *Adv. Eng. Mater.* **2005**, *7*, 401.

[14] N. Carneiro, A. P. Souto, E. Silva, A. Marimba, B. Tena, H. Ferriera, V. Magalhaes, *Color. Technol.* **2001**, *117*, 298.

[15] M. Prabaharan, N. Carneiro, *Indian J. Fibre Text.* **2005**, *30*, 68.

[16] E. Özdogan, R. Saber, H. Ayhan, N. Seventekin, *Color. Technol.* **2002**, *118*, 100.

[17] H. -U. Poll, S. Schreiter, *Melliand Textilberichte* **1998**, *6*, 466.

[18] A. M. Sarmadi, T. H. Ying, F. Denes, *Text. Res. J.* **1993**, *63 (12)*, 697.

[19] D. Hegemann, *Indian J. Fibre Text.* **2006**, *31*, 99.

[20] M. Keller, A. Ritter, P. Reimann, V. Thommen, A. Fischer, D. Hegemann, *Surf. Coat. Technol.* **2005**, *200*, 1045.

[21] E. Kissa, *Text. Res. J.* **1996**, *66 (10)*, 660.

[22] H. Ito, Y. Muraoka, *Text. Res. J.* **1993**, *63 (7)*, 414.

[23] M. Pociute, B. Lehmann, A. Vitkauskas, *Mater. Sci.* (Medziagotyra) **2003**, *9 (4)*, 410.

[24] Y. K. Kamath, S. B. Hornby, H. -D. Weigmann, M. F. Wilde, *Text. Res. J.* **1994,** *64 (1)*, 33.

[25] D. Hegemann, U. Schütz, *Thin Solid Films* **2005**, *491*, 96.

[26] C. Oehr, D. Hegemann, M. Müller, U. Vohrer, M. Storr, in: *Plasma Process. Polym.*, eds. R. d'Agostino, P. Favia, C. Oehr, M. R. Wertheimer, Wiley-VCH, Weinheim, **2005**, 309.

[27] G. Placinta, F. Arefi-Khonsari, M. Gheorghiu, J. Amouroux, G. Popa, *J. Appl. Polym. Sci.* **1997**, *66*, 1367.

[28] M. M. Hossain, D. Hegemann, P. Chabrecek, A. S. Herrmann, *J. Appl. Polym. Sci.,* **2006**, *102*, 1452.

[29] T. Öktem, N. Seventekin, H. Ayhan, E. Piskin, *Indian J. Fibre Text.* **2002**, *27 (6)*, 161.

[30] H. Höcker, *Pure Appl. Chem.* **2002**, *74 (3)*, 423.

[31] C. Jie-Rong, W. Xue-Yan, W. Tomiji, *J. Appl. Polym. Sci.* **1999**, *72*, 1327.

[32] N. Inagaki, K. Narushima, S. K. Lim, *J. Appl. Polym. Sci.* **2003**, *89*, 96.

[33] R. W. Paynter, *Surf. Interface Anal.* **2000**, *29*, 56.

[34] V. A. Volkov, B. V. Bulushev, A. A. Ageev, *Colloid J.* **2003**, *65*, 523.

5. Influence of Reactive Gases Added During Plasma Polymerization

5.1. Introduction

Plasma polymerization refers to the deposition of thin films showing a broad range of properties using different monomer gases under varying plasma conditions such as pressure, power input, flow, temperature, and plasma potentials.[1] Specifically molecular fragmentation in the plasma phase and crosslinking during film growth determine the film's properties ranging from soft, swellable or highly functional coatings to hard, crosslinked films. Different substrates and geometries such as textiles, fibers, and membranes etc. can be treated with high plasma penetration. A broad parameter range is available, but the distinct influence of reactor geometry complicates the process optimization and scaling-up.[2] Therefore, a deeper understanding of process relevant correlations on a macroscopic scale is also desirable.

Although plasma polymerization is applicable to pure monomer discharges, carrier gases are often added to increase the homogeneity and stability of the plasma and to enhance the possible ways of monomer excitation. Thus, carrier gases such as Ar, He etc. can be considered as energy-carrying gases. Furthermore, reactive gases can be added, which may support the deposition and formation of additional functional groups (acting as energy-consuming co-monomers) or yield etching effects.

Yasuda extensively studied the influence of non-polymerizable gases on plasma polymerization. He showed that Ar and H_2 gases used in hydrocarbon plasmas had only a minor effect on deposition rates, whereas N_2 had a strong influence on deposition rates.[3] However, according to the existing literature, there are still some gaps concerning the contribution of additional (carrier or reactive) gases to plasma polymerization. Some authors have reported that the deposition rate is enhanced when non-polymerizable gases are added to monomer gases during polymerization,[4,5] while others found a less distinct tendency.[6-8] Therefore, the research reported in this chapter is an attempt to clarify the situation by taking a macroscopic approach to plasma polymerization.

5.1.1. Plasma Polymerization and Growth Mechanism

Numerous references[9-12] suggest that radicals predominantly take part in the deposition process (radical-dominated plasma) at moderate power, though a plasma always contains also neutral gas particles, excited states, ions, electrons, and UV beside radicals. Beck et al.[13] reported that cationic species can have a strong influence on plasma polymerization. Plasma deposition characteristics and plasma process parameters were examined by numerical modeling techniques.[10,14] Reactor geometry and gas flow types have been found to be important factors which also affect plasma polymerization.[15,16] However, radical-dominated plasma polymerization mechanisms remain a subject of discussion due to the following major reasons: i) the radical concentration in a non-equilibrium hydrocarbon plasma is higher than the ion concentration. This is because less energy is required for dissociation compared to the ionization processes,[11] and ii) high radical concentrations in the plasma polymers, which are very reactive to the hydrocarbon monomers.[9]

Figure 1 describes the initiation and propagation of radical-induced plasma polymerization, which is mainly based on hydrogen abstraction from monomer and/or by bond opening.[9,17] Interface reactions between the plasma and plasma-polymer were ignored. These reactions are only significant at very high energy levels.[18] When hydrocarbon monomer molecules enter the plasma zone, they collide with free electrons, and plasma polymerization is initiated. Surface free radicals can be formed by UV, electrons and positive ions on the growing polymer surfaces. Reactive intermediates such as hydrogen atoms, free radicals (mono-/bi-radicals) are generated by initiation reactions. It is clear from the reaction-mechanisms (reaction 1 in Figure 1) that atomic hydrogen does not directly take part in the film deposition. However, it strongly influences the deposition reactions. Atomic hydrogen is considered to determine the hydrocarbon layer growth by transformation of higher hydrides into mono-hydrides yielding reduced hydrogen incorporation in a-C:H films and radical sites.[19] Moreover, hydrogen atoms etch the surface chemically, resulting in lower deposition rates.[20] Hydrogen detachment and/or the bond opening of unsaturated monomers contribute significantly to the formation of mono-/bi-radicals, respectively. These radicals can further react with monomer molecules or they can recombine with each other. Consequently, large molecules can be formed on the growing plasma polymer surface (reactions 2-6 in Figure 1). In addition, reactions between plasma species (ions, energetic particles, radicals) and the growing film yield crosslinking reactions. Thus, plasma polymers do not resemble conventional polymers showing merely short repeating units.

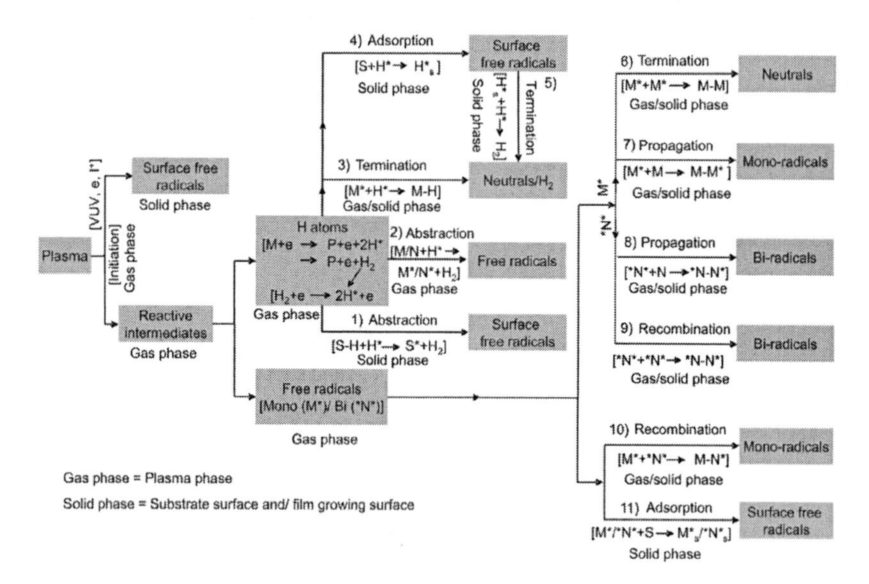

Figure 1. Initiation and propagation of radical-induced plasma polymerization.[14]

5.1.2. Evaluation of Deposition Rates

As it is well-known, plasma polymerization – i.e. radical-promoted – processes are governed by the composite parameter power input per gas flow W/F, which relies on the concept of macroscopic kinetics.[21] This statistical approach considers the energy invested per particle within the active plasma zone yielding an excited state such as a radical and subsequent recombination in a passive zone yielding a stable product such as a deposition. A monomer-dependent activation energy E_a* can be derived by evaluating the mass deposition rates R_m[2,22,23] considering that

$$\frac{R_m}{F} = G\exp\left(-\frac{E_a}{W/F}\right) \tag{5.1}$$

with a reactor-dependent geometrical factor G. An Arrhenius-type plot of Equation (5.1) by variation of W/F parameters yields a straight line, where its slope corresponds to E_a (Figure 2). While Equation (5.1) agrees very well with experimentally found deposition rates for many different monomers at energy inputs around E_a, some deviations might be obtained at low power inputs due to different precursors or oligomers participating in film growth and at high power inputs indicating ion-assisted or temperature effects. Hence, different effects can be observed using the evaluation of deposition rates.

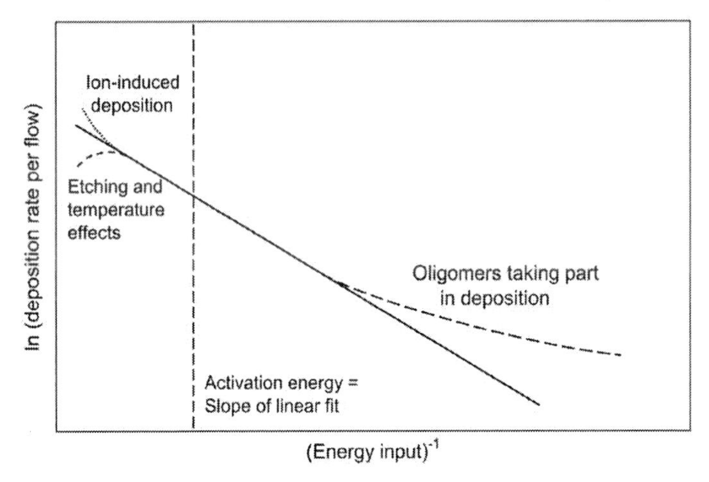

Figure 2. Schematic Arrhenius-type plot of the mass deposition rate per flow R_m/F depending on the inverse energy input $(W/F)^{-1}$. The linear part around the activation energy is described by Equation (5.1).

* E_a represents an apparent activation energy (in J/cm^3) which is related to the dissociation of film-forming radicals.

It was recently[2] proved that, for example, the deposition of plasma polymerized hexamethyldisiloxane (pp-HMDSO) films follows Equation (5.1) over a wide parameter range using different reactor types and plasma sources.[2] Therefore, HMDSO discharges are radical-dominated plasmas, since the mass deposition rate is based on the mean energy consumed per monomer molecule within the active plasma zone. The slope of the resulting curve of Equation (5.1) is thus independent of ion-induced effects during film growth. Moreover, using an optimized reactor or regarding a reactor-dependent geometrical factor, Hegemann et al. showed that the activation energy derived from Equation (5.1) is indeed related to the energy required per molecule to initiate plasma polymerization.[24,25] Using a symmetrical reactor with low ion bombardment, pp-HMDSO coatings were deposited at an energy input around E_a which strongly resemble polydimethylsiloxane.[23]
In addition to macroscopic kinetics, the energy balance at the substrate surface during plasma polymerization was also used to calculate the deposition rate based on activation reactions, transport processes in the plasma volume and reactions at the surface.[26] Both the influence of energy influx and resulting thermal conditions can thus be considered. By means of the derived adsorption model the growth rate R_{pol} of the polymer film is given by

$$R_{pol} = \sum_{\mu} \sigma_{\mu} n_{\mu} j_{x} \tag{5.2}$$

with the reaction cross section σ_{μ} for the incorporation of monomer molecules of kind μ into the thin film under the action of collisions with an energetic particle flux j_x and the actual surface concentration of monomer molecules n_{μ}.[27] Thus, the deposition rate increases with the particle flux density to the substrate, which is controlled by the energy input into the plasma. Moreover, the model showed that at relatively low substrate temperatures the growth rate depends only on the monomer supply and is independent of the temperature, while at elevated temperatures the deposition rate decreases due to a reduced residence time of monomer molecules at the substrate surface.[27]
Regarding the adsorption model it can be inferred that the action of energetic particles during plasma polymerization is inherently contained within the macroscopic kinetics model. This means that an increased energy input W/F both increases fragmentation and ion-supported deposition, where the latter influences merely the geometrical factor G and thus the absolute deposition rate, but not the monomer-dependent activation energy. This means that the same plasma chemistry is taking place within the linear range of Equation (5.1) contributing to film deposition. The adsorption model according to Equation (5.2) also indicates that the contribution of different monomer molecules or reactive particles should be added together to yield the corresponding reaction cross sections. Thus, the gas flows should be added together during plasma polymerization as

$$F = F_m + aF_c \tag{5.3}$$

with the monomer flow F_m, the carrier or reactive gas flow F_c and a weighting factor a (flow factor).[28] In the following it will be proven that the reaction parameter W/F can still be used to evaluate deposition rates with the modified flow according to Equation (5.3). Therefore, own experiments were examined and the data reported in the literature were discussed.

5.2. Experimental

As a prerequisite for the presented concept, an appropriate plasma reactor with well defined characteristics such as a continuous and reproducible change in plasma conditions with external parameters is indispensable. Thus, a clearly arranged set-up designed to avoid dead spaces and to provide a gas flow directed towards the deposition area might be beneficial. Therefore, symmetric plasma reactors were used, where the RF-driven electrode (13.56 MHz) was comparable in size to the counter electrode and the whole space between the electrodes was filled by the plasma. This set-up provided a well-defined active plasma zone, where the gas particles were excited. For the deposition of SiO_x films from O_2/HMDSO mixtures a plane parallel configuration with an electrode area of 1150 cm^2 and electrode distance of 8.5 cm was used, which is described elsewhere.[28] The working pressure was in the range between 10 and 50 Pa. To obtain a-CN_x:H films, acetylene and ammonia were mixed at a pressure of 10 Pa using a web coater.[25,29] Samples were fixed on a rotatable, cylindrical drum (59 cm in diameter, 70 cm in length) acting as RF electrode, while the vacuum chamber represented the counter electrode at a distance of 7 cm. Mass deposition rates (in mass gain per area and time) were obtained by weighing glass slides before and after deposition.
In order to obtain the same mass deposition rates as those cited in the literature data, the deposition rates given in film thickness per unit of time would have to be multiplied by the corresponding film density, which is not always reported. However, it is known that for example a-C:H films show a strong correlation between film density, hydrogen content and refractive index independent of the reactor or plasma source used.[30] Therefore, densities were used to observe the comparable plasma coatings. Hence, there might be a deviation regarding the absolute mass deposition rate, however, the trend in the slope of the Arrhenius-type plots is unaffected.

5.3. Result and Discussion

In order to investigate the validity of Equations (5.1) and (5.3) plasma polymerization was considered with different monomers and non-polymerizable gases.

5.3.1. Oxygen Added to HMDSO Discharges

Gas mixtures of oxygen and HMDSO were examined over a broad parameter range. Regarding only the monomer flow F_m for the reaction parameter *W/F* it was found that the deposition rates follow Equation (5.1) for each gas ratio with increasing slope, i.e. activation energy, at increasing O_2/HMDSO ratio (Figure 3, left).[28] Introducing the flow factor *a* and considering the summed flow according to Equation (5.3) for the reaction parameter *W/F*, a merging of the different slopes at $a = 0.6$ (Figure 3, right) was observed. Using Equations (5.1) and (5.3) the contribution of the oxygen addition can be further analyzed. Regarding the mass deposition rate R_m for increasing O_2/HMDSO ratios at the corresponding activation energy compared to the deposition rate for pure HMDSO at the same monomer flow and power input, the addition of oxygen yields higher deposition rates (Figure 4). At the same time the oxygen incorporation into the SiOC:H films is increased and the carbon content is reduced.

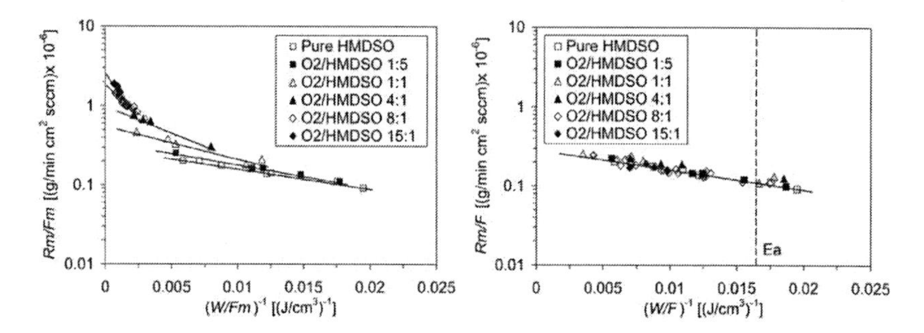

Figure 3. Mass deposition rates of O_2/HMDSO discharges plotted against $(W/F)^{-1}$. The different curves obtained for different gas ratios regarding only the monomer flow F_m (left) merge into one curve representing one activation energy by considering the generalized total flow $F = F_m + a\ F_c$ for a = 0.6 (right).

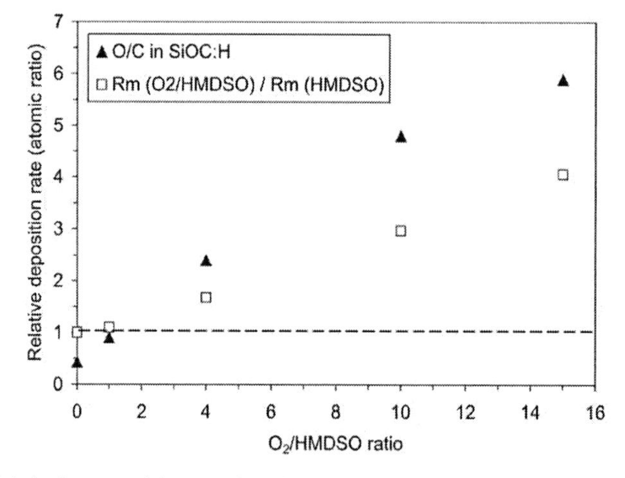

Figure 4. Relative increase of the mass deposition rates, as well as the O/C ratios in the SiOC:H films deposited under similar plasma conditions with increasing O_2/HMDSO ratios.

Thus, the high flow factor of 0.6 for O_2 added to HMDSO indicates that oxygen directly contributes to the polymerization process, whereby the O content in the films can be enhanced from around $SiOC_2$ (PDMS-like) towards SiO_2 (quartz-like). The strong contribution of oxygen to the film growth is also supported by the finding that even a small addition of O_2 noticeably influences the discharge conditions, the plasma components as well as the chemical film composition.[4,28,31]

Fang et al. noticed the strongest influence of oxygen on HMDSO discharges followed by Ar, N_2, He, and H_2 as carrier gases using an inductively coupled plasma at 27 Pa, with constant monomer and carrier gas flow and an energy input around the activation energy, which can be derived by evaluation of their deposition rates.[4] Regarding Equations (5.1) and (5.3) it can be assumed that Ar, N_2, He, and H_2 (in this order) show lower flow factors when compared to O_2 with a = 0.6. While they also found an increased oxygen incorporation by using O_2, they obtained rather

PDMS-like coatings for the other carrier gases examined. Especially, He and H_2 have merely a small influence on HMDSO plasma polymerization.[32]

5.3.2. N-containing Gases Added to C_xH_y Discharges

The influence of a potential etching gas on the plasma polymerization of acetylene was investigated using the web coater. The use of pure C_2H_2 yield the deposition of amorphous hydrocarbon coatings (a-C:H), the properties of which are extensively described in the literature.[30,33-35] Adding a nitrogen source during the plasma polymerization CN_x:H coatings can be deposited, which is also a well known process. Generally, it is believed that the amount of nitrogen that can be incorporated into the a-C:H films is limited to 30-40 at%, which is accompanied by a strong reduction in the deposition rate, at least if hard coatings are to be attained.[36,37] This finding is ascribed to a transition between the growth and erosion regimes, i.e. etching effects of nitrogen during film growth, which is observed for noticeably higher N than C concentrations in the plasma. Only in the case of lower energy inputs, polymer-like films with higher amounts of nitrogen can be obtained.[37] Therefore, the mass deposition rates for NH_3/C_2H_2 gas mixtures of 1:1, 2:1, 3:1, and 5:1 in comparison to pure acetylene discharges at a pressure of 10 Pa were examined. Using the monomer flow F_m for the specific energy W/F again different slopes are obtained (Figure 5, left), which can be adjusted by fitting the flow factor *a*. (Figure 5, right). At a = 0.5, the same activation energy was obtained for all gas ratios.

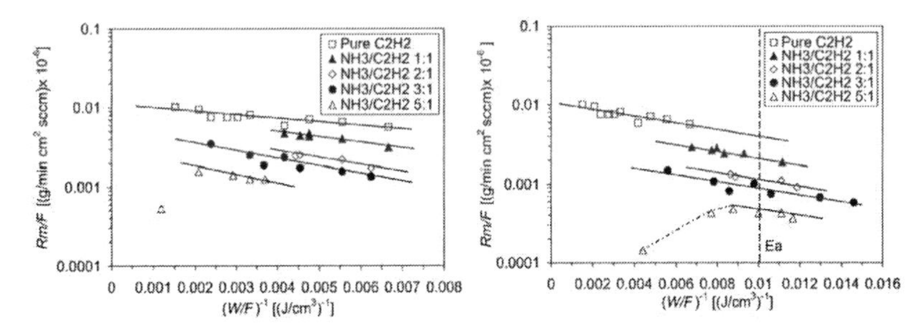

Figure 5. Mass deposition rates of NH_3/C_2H_2 discharges plotted against $(W/F)^{-1}$. The curves obtained for different gas ratios regarding only the monomer flow F_m (left) reveal the same slope (right, see text for details).

Unlike the O_2/HMDSO deposition, this time the absolute deposition rate decreased with increasing NH_3/C_2H_2 ratio. A comparison of the deposition rates for different NH_3/C_2H_2 ratios at similar plasma conditions (energy input corresponding to the activation energy) revealed a relative decrease in deposition rate for increasing ammonia content (Figure 6), which was ascribed to chemical etching of physisorbed hydrocarbon precursors during the film growth by N-containing species.[38] At the same time, the amount of N incorporated into the a-CN_x:H films increased. At higher NH_3/C_2H_2 ratios (>3) and higher energy inputs (>E_a) an additional etching effect occurred that strongly decreases the deposition rate. Here, a transition from the radical-induced deposition (determined by the activation energy) to the erosion regime (determined by ion bombardment) can be noticed, which is no longer described by Equation (5.1).

Waldman et al. reported on the ethylene/ammonia plasma polymer deposition in a comprehensive study.[39] They looked at inductively coupled RF discharges with 100, 75, 50, and 25% C_2H_4 diluted in NH_3 at a total flow of 40 sccm and a rather high pressure of 480 Pa. For the pure ethylene plasma they obtained a high amount of powder production, which might have influenced (reduced) the observed deposition rate.[40] Moreover, they found higher deposition rates for higher NH_3 flows. Probably due to the high pressure and the floating conditions inside their tube reactor, no etching conditions were obtained. Incorporation of nitrogen yielded instead a slightly increased film thickness with increasing NH_3/C_2H_4 ratio.[3] However, interpretation of their reported deposition rates again reveals the validity of the presented concept regarding the flow $F = F_m + a\ F_c$, where $a = 0.5$ for reasonably assumed densities of the deposited films (Figure 7). At energies far below the activation energy, enhanced deposition rates were obtained due to oligomers taking part in the film growth. On the other hand, the energy input around the apparent activation energy was not high enough to obtain etching conditions. Waldman obtained CN_x:H films with an N content around 20 at% for a NH_3/C_2H_4 ratio of 3:1.

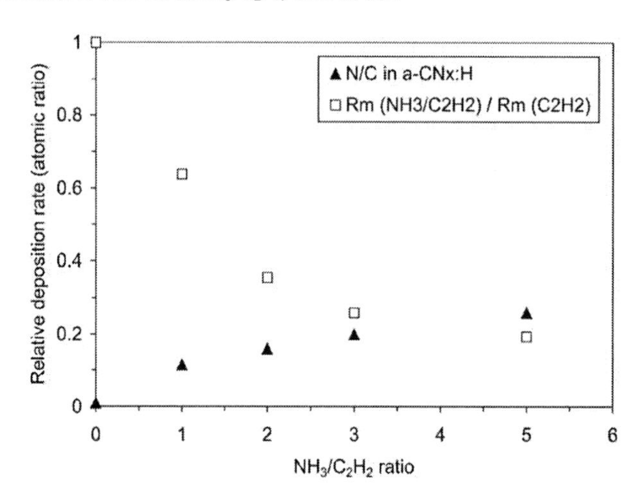

Figure 6. Relative decrease of the mass deposition rate with increasing NH_3/C_2H_2 ratio at comparable conditions. At the same time the nitrogen to carbon ratio in the a-CN_x:H films increases.

Mutsukura and Akita investigated pure CH_4 discharges and compared them to 10% CH_4 diluted in N_2 using an asymmetric RF plasma reactor at a pressure of 130 Pa.[41,42] While the deposition rate within the pure CH_4 discharge increased with increasing RF power (at constant flow), they found that the N_2/CH_4 mixture led to a decrease in deposition rate above a certain power input, which was attributed to etching/sputtering processes. A thorough analysis of their reported deposition rates according to Equation (5.1) showed that both the deposition rate of pure CH_4 and the CH_4/N_2 mixture could be described by the flow $F = F_m + a\ F_c$ with $a = 0.35$ as long as the energy input was below the apparent activation energy (Figure 8). Density values between 1.1 and 1.4 g/cm^3 were assumed for CH_4 and 1.2 to 1.8 cm^3 for CH_4/N_2

according to the related power per monomer flow compared with the experiments presented here. Hence, the excess of nitrogen-containing gases added to hydrocarbons yielded strong sputtering conditions when the energy input was above the activation energy. Energetic N_2^+ ion bombardment seems to be responsible for this finding.[43] Moreover, the chemical composition of the a-CN_x:H films were found to change remarkably with ion bombardment.[44]

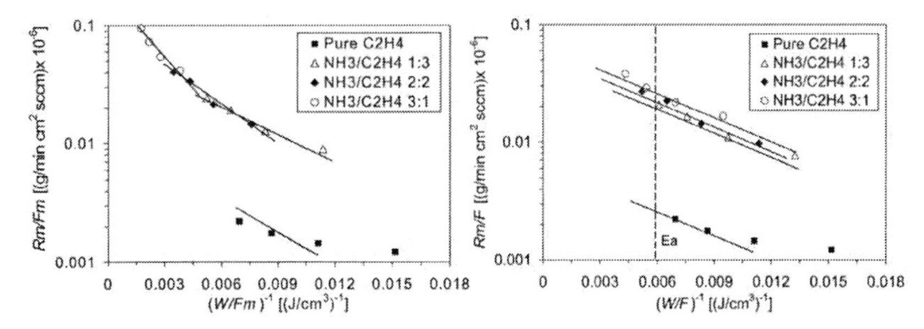

Figure 7. Mass deposition rates of NH_3/C_2H_4 discharges plotted against $(W/F)^{-1}$. The curves obtained for different gas ratios regarding only the monomer flow F_m (left) reveal the same slope, i.e. the same activation energy, by considering the generalized total flow $F = F_m + a\, F_c$ for $a = 0.5$ (right). Data taken from.[39]

At lower nitrogen admixtures to methane (up to 50%), no transition to strong sputtering conditions was observed by Freire using a capacitively coupled RF discharge at 10 Pa, a bias voltage of -350 V (30 W), and a total flow of 3 sccm.[45] However, he observed a decrease in the deposition rate with increasing N_2 fractions. This behavior can be well understood by regarding the reaction parameter W/F with $F = F_m + a\, F_c$ at a constant total flow and an energy input above E_a as well as the discussed competition between aggregation and erosion during film growth, while no N_2^+ sputtering mechanism has to be supposed. At N_2/CH_4 (and NH_3/C_2H_4) ratios greater than one a strong reduction in deposition rate was reported indicating the onset of sputtering.[46]

N_2 added to CH_4 merely yielded a small incorporation of nitrogen into the a-C:H films (10-15 at%), which resulted in reduced internal stresses by an increase of the number/size of sp^2-carbon clusters.[45] However, N_2 addition retained (or slightly increases) the amount of CH_3 radicals and led to the formation of CN precursors in the gas phase,[47] which influenced the polymerization rate causing the rather high factor of 0.35. Interestingly, high ratios of N_2 or NH_3 to hydrocarbon gases were also considered for the deposition of carbon nanotube (CNT) films (at high energy inputs).[48,49]

5.3.3. Inert Gases Added to C_xH_y Discharges

When adding Ar or H_2 to CH_4 discharges up to a ratio of 2:1 no distinct influence of the non-polymerizable gases to the deposition rate or to film properties were found. Similar results were obtained by Tomasella et al. by adding Ar or He up to a ratio of 2:1 to a methane RF plasma at a low pressure of 1 Pa[50] and by Schulz-von der Ganthen et al. by admixtures of Ar or H_2 up to 4:1 at a higher pressure of 100 Pa.[51]

Hence, a noticeably smaller flow factor can be expected for inert gases compared to reactive gases such as O_2, NH_3, and N_2. However, at higher admixtures an influence on the degree of dissociation of CH_4 was observed. Jacobsohn et al. compared pure methane discharges to highly diluted Ar/CH_4 discharges (49:1).[52] They detected no remarkable differences in film properties, however, argon was found to influence the deposition rate. An asymmetric, cylindrical RF reactor was used at 13 Pa. By analyzing their reported deposition rates, two curves were obtained for the two series of pure methane and Ar/CH_4 mixture (Figure 9). While for an energy input well below the activation energy, a higher deposition rate compared to the linear fit was obtained, presumably due to oligomers taking part in the film growth, elevated deposition rates were also found for high energy inputs perhaps due to ion-induced nucleation sites at the film surface. Using the same approach as for the addition of reactive gases to monomers, both curves merged into one, when the factor a (in $F = F_m + a F_c$) is chosen to be 0.05 (Figure 9). Thus, only a small part of the energy input that contributed to film deposition was consumed by the carrier gas, and for small argon admixtures no influence was observed. Looking at data reported for Ar/C_2H_2 mixtures (1:3, 1:1, and 3:1),[38] a flow factor a of about 0.1 was derived. Using Equation (5.1) and (5.3) plasma polymerization can thus be investigated helping also to identify regimes of chemical etching/physical sputtering and clarify controversial discussions found on the literature.[1]

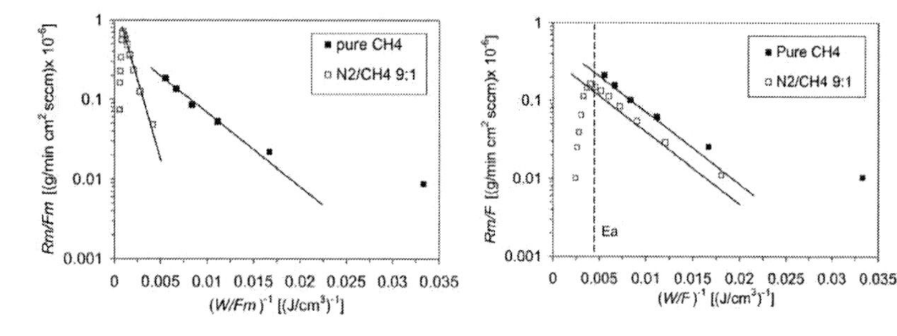

Figure 8. Mass deposition rates of N_2/CH_4 discharges plotted against $(W/F)^{-1}$. The curves obtained for different gas ratios regarding only the monomer flow F_m (left) reveal the same slope, i.e. the same activation energy, by considering the generalized total flow $F = F_m + a F_c$ for a = 0.35 (right). Data taken from.[41,42]

Hence, plasma polymerization is an inherently plasma chemical process concerning the dependence of the amount of deposit from the vapor phase on the reaction parameter W/F. Nevertheless, physical effects by energetic particles such as ions[30,53] or atom beams[54] influence the film characteristics, mainly by determining the density and the internal stresses during the polymerization process at the substrate surface. To infer possible sputtering effects the obtained deposition rates should be plotted according to Equation (5.1), whereby deviations at higher energy inputs might indicate additional, mainly ion-induced effects.

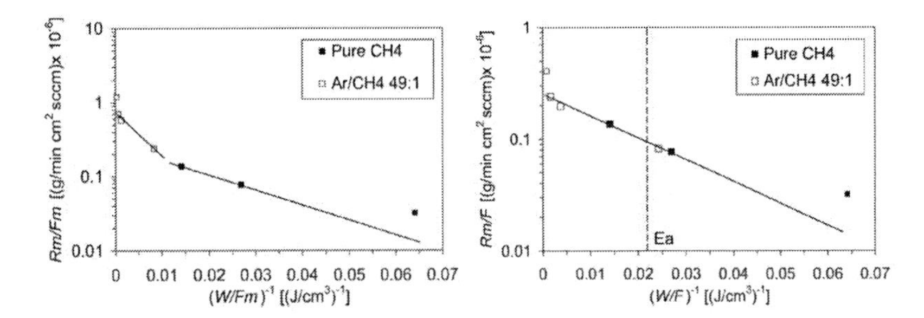

Figure 9. Mass deposition rates of a pure CH_4 discharge and a highly diluted Ar/CH_4 discharge plotted against $(W/F)^{-1}$. The obtained curves regarding only the monomer flow F_m (left) reveal the same slope, i.e. the same activation energy, by considering the generalized total flow $F = F_m + a\ F_c$ for a = 0.05 (right). Elevated deposition rates were obtained for energies well above or below the activation energy E_a. Data taken from.[52]

5.4. Conclusion

The influence of non-polymerizable gases added to monomer discharges was investigated by evaluating the mass deposition rates. This concept is based on macroscopic kinetics and on the adsorption model showing that plasma polymerization is determined by the reaction parameter power input per gas flow W/F, where the flow is a sum of monomer and carrier gas flow $F = F_m + a\ F_c$ regarding a flow factor a. The carrier gas might be either an inert or a reactive gas. This concept was proved for O_2/HMDSO and NH_3/C_2H_2 discharges. Moreover, the data on deposition rates given in the literature (N_2, NH_3, Ar etc. added to hydrocarbon discharges) were analyzed, which were found to support this novel approach. The slope of the deposition rates in an Arrhenius-type plot indicates the activation energy of the corresponding process. By introduction of the flow factor a different activation energies obtained by varying the gas ratios can be adjusted yielding a single activation energy for each specific polymerizable monomer used. The change in the deposition rate, i.e. the activation energy, depending on the reaction parameter W/F is thus determined by plasma-chemical effects within the active plasma zone (formation of radicals). Since the specific energy W/F corresponds to the energy invested per particle, the flow factor indicates the distribution of the energy consumed by polymerizable and non-polymerizable particles.
Nitrogen in the plasma state is chemically reactive and should be considered a co-monomer of the plasma polymerization, although it does not polymerize alone in plasma.[3] This behavior is reflected in the flow factor a of 0.35-0.5 (added to hydrocarbons). Oxygen, which is an important co-monomer in the deposition of organosilicones, even shows a flow factor of 0.6, whereas inert gases such as argon or helium were found to have lower flow factors. It should be noted that the absolute deposition rate might be influenced by the addition of non-polymerizable gases due to chemical etching effects. Deviations of the obtained deposition rates from those shown in Equation (5.1) at higher energy inputs can be a further indication of ion-induced effects, e.g. sputtering. However, the results presented here show that the main influence of non-polymerizable gases on plasma polymerization can be

explained by simply considering the modified reaction parameter W/F with $F = F_m + a F_c$. Using this modified flow the change obtained in the deposition rate by adding non-polymerizable gases is shown to depend on the energy input (with respect to the activation energy), on the gas ratio, and on the absolute monomer flow (constant or varying), which can be related to the energy- and monomer-deficient plasma polymerization regimes.

This statistical, macroscopic approach makes it possible to optimize and scale-up plasma polymerization processes as it regards the plasma chemistry determined by the activation energy.[12] On the contrary, the process transfer of plasma activation and etching processes with non-polymerizable gases between various reactors was found to depend on the energy flux to the substrate, which is mainly determined by the sheath voltage and the plasma power density (power per area) controlling the ionization degree.[27,55] Thus, a general difference between plasma polymerization and etching/sputtering processes should be considered. Using macroscopic kinetics the radical-dominated polymerization regime can be identified, while deviations at higher energy inputs or higher gas ratios of non-polymerizable to polymerizable gas indicate the influence of energetic particles. This findings were used to investigate hydrophobic (HMDSO-derived) and nanoporous C-H-(N,O) coatings as shown in the following chapter.

5.5. References

[1] D. Hegemann, M. M. Hossain, *Plasma Process. Polym.* **2005**, *2*, 554.
[2] D. Hegemann, C. Oehr, A. Fischer, *J. Vac. Sci. Technol. A* **2005**, *23*, 5.
[3] H. Yasuda, Plasma Polymerization, Academic Press, Orlando, **1985**.
[4] J. Fang, H. Chen, X. Yu, *J. Appl. Polym. Sci.* **2001**, *80*, 1434.
[5] E. Braca, J. M. Kenny, D. Korzec, J. Engemann, *Thin Solid Films* **2001**, *394*, 30.
[6] A. Barranco, J. Cotrino, F. Yubero, J. P. Espinós, J. Benítez, C. Clerc, A. R. González-Elipe, *Thin Solid Films* **2001**, *401*, 150.
[7] E. Bapin, P. Rudolf von Rohr, *Surf. Coat. Technol.* **2001**, *142-144*, 649.
[8] S. Sahli, S. Rebiai, P. Raynaud, Y. Segui, A. Zenasni, S. Mouissat, *Plasmas Polym.* **2002**, *7*, 327.
[9] J. M. Tibbit, R. Jensen, A. T. Bell, M. Shen, *Macromolecules* **1977**, *10 (3)*, 647.
[10] L. V. Shepsis, P. D. Pedrow, R. Mahalingam, M. A. Osman, *Thin Solid Films,* **2001**, *385*, 11.
[11] I. Chen, *Thin Solid Films* **1983**, *101*, 41.
[12] D. Hegemann, A. Fischer, *J. Vac. Sci. Technol. A* **2005**, *23 (1)*, 5.
[13] A. J. Beck, R. D. Short, *J. Vac. Sci. Technol. A* **1998**, *16 (5)*, 3131.
[14] R. J. Buss, *J. Appl. Polym. Sci.* **1986**, *59 (8)*, 2977.
[15] D. Hegemann, *Indian J. Fibre Text.* **2006**, *31*, 99.
[16] D. Hegemann, D. J. Balazs, M. Amberg, A. Fischer, *Proc. 17th Int. Symp. on Plasma Chemistry*, August 07-12, **2005**, Toronto/Canada.
[17] H. Yasuda, *"Luminous Chemical Vapor Deposition and Interface Engineering"*, Marcel Dekker, USA, **2005**, pp. 64-65.
[18] M. Horie, *J. Vac. Sci. Technol. A* **1995**, *13 (5)*, 2490.
[19] A. von Keudell, *Thin Solid Films* **2002**, *402*, 1.
[20] T. Mieno, T. Shoji, K. Kadota, *Jpn. J. Appl. Phys.* **1992**, *31*, 1879.
[21] A. Rutscher, H. -E. Wagner, *Plasma Sources Sci. Technol.* **1993**, *2*, 279.
[22] Y. S. Yeh, I. N. Shyy, H. Yasuda, *J. Appl. Polym. Sci.: Appl. Polym. Symp.* **1988**, *42*, 1.
[23] D. Hegemann, H. Brunner, C. Oehr, *Plasmas Polym.* **2001**, *6*, 221.
[24] D. Hegemann, U. Schütz, A. Fischer, *Surf. Coat. Technol.* **2005**, *200*, 458.
[25] D. Hegemann, in: Plasma Polymers and Related Materials, ed. M. Mutlu, Univ. Ankara, **2005**.
[26] H. Deutsch, M. Schmidt, *Beitr. Plasmaphys.* **1981**, *21*, 279.
[27] H. Kersten, H. Deutsch, H. Steffen, G.M.W. Kroesen, R. Hippler, *Vacuum* **2001**, *63*, 385.
[28] D. Hegemann, H. Brunner, C. Oehr, *Surf. Coat. Technol.* **2003**, *174-175*, 253.
[29] D. Hegemann, A. Fischer, *Proc. Int. Textile Congress*, October 18-20, **2004**, Barcelona/Spain.
[30] T. Schwarz-Selinger, A. von Keudell, W. Jacob, *J. Appl. Phys.* **1999**, *86*, 3988.
[31] A. Granier, M. Vervloet, K. Aumaille, C. Vallée, *Plasma Sources Sci. Technol.* **2003,** *12*, 89.
[32] S. Gaur, G. Vergason, *Proc. SVC 43rd Ann. Techn. Conf.*, April 15-20, **2000**, Denver, 267.
[33] D. R. McKenzie, R. C. McPhedran, N. Savvides, D. J. H. Cockayne, *Thin Solid Films* **1983**, *108*, 247.

[34] S. Vasquez-Borucki, W. Jacob, C. A. Achete, *Diam. Relat. Mater.* **2000**, *9*, 1971.
[35] R. A. DiFelice, J. G. Dillard, D. Yang, Int. *J. Adhes. Sci.* **2005**, *25*, 342.
[36] F. L. Freire Jr., G. Mariotto, R. S. Brusa, A. Zecca, C. A. Achete, *Diam. Relat. Mater.* **1995**, *4*, 499.
[37] P. Hammer, F. Alvarez, *Thin Solid Films* **2001**, *398-399*, 116.
[38] Y. H. Cheng, Y. P. Wu, J. G. Chen, X. L. Qiao, C. S. Xie, B. K. Tay, S. P. Lau, X. Shi, *Surf. Coat. Technol.* **2000**, *135*, 27.
[39] D. A. Waldman, Y. L. Zou, A. N. Netravali, *J. Adhes. Sci. Technol.* **1995**, *9*, 1475.
[40] J. Berndt, S. Hong, E. Kovacevic, I. Stefanovic, J. Winter, *Vacuum* **2003**, *71*, 377.
[41] N. Mutsukura, K. Akita, *Diam. Relat. Mater.* **2000**, *9*, 761.
[42] N. Mutsukura, K. Akita, *Thin Solid Films* **1999**, *349*, 115.
[43] K. J. Clay, S. P. Speakman, G. A. J. Amaratunga, S. R. P. Silva, *J. Appl. Phys.* **1996**, *79*, 7227.
[44] N. Mutsukura, Y. Daigo, *Diam. Relat. Mater.* **2003**, *12*, 2057.
[45] F. L. Freire Jr., *J. Non-Crystalline Solids* **2002**, *304*, 251.
[46] L. G. Jacobsohn, F. L. Freire, Jr., D. F. Franceschini, M. M. Lacerda, G. Mariotto, *J. Vac. Sci. Technol. A* **1999**, *17*, 545.
[47] F. Hempel, J. Röpcke, A. Pipa, P. B. Davies, *Mol. Phys.* **2003**, *101*, 589.
[48] P. Gröning, P. Ruffieux, L. Schlapbach, O. Gröning, *Adv. Eng. Mater.* **2003**, *5*, 541.
[49] S. Hofman, B. Kleinsorge, C. Ducati, J. Robertson, *New J. Phys.* **2003**, *5*, 1531.
[50] E. Tomasella, C. Meunier, S. Mikhailov, *Surf. Coat. Technol.* **2001**, *141*, 286.
[51] V. Schulz-von der Ganthen, J. Röpcke, T. Gans, M. Käning, C. Lukas, H. F. Döbele, *Plasma Sources Sci. T.* **2001**, *10*, 530.
[52] L. G. Jacobsohn, G. Capote, N. C. Cruz, A. R. Zanatta, F. L. Freire Jr., *Thin Solid Films* **2002**, *419*, 46.
[53] Y. Lifshitz, G. D. Lempert, S. Rotter, I. Avigal, C. Uzan-Saguy, R. Kalish, J. Kulik, D. Marton, J. W. Rabalais, *Diam. Relat. Mater.* **1994**, *3*, 542.
[54] O. S. Panwar, D. Sarangi, S. Kumar, P. N. Dixit, R. Bhattacharyya, *J. Vac. Sci. Technol. A* **1995**, *13*, 2519.
[55] J. A. G. Baggerman, R. J. Visser, E. J. H. Collart, *J. Appl. Phys.* **1994**, *75*, 758.

6. Dyeability of Nanoporous Coatings Using NH_3/C_2H_2 Discharges

6.1. Introduction

Polyester (PES) fabrics are the most widely used synthetic fiber in domestic, biomedical, and industrial applications.[1-4] PES fabrics possess very good physical and chemical properties such as high tensile strength, resistance to stretching and shrinking, abrasion resistance, resistance to heat, strength retention, and resistance to sunlight, oxidizing agents, acids, alkalis, bleaches etc.[4] In addition to their common uses in textiles and clothing, they have great importance in technical textiles, home textiles, automobile industry, as well as medical textiles.

PES fabrics are broadly used in blends with natural fibers. PES fabrics are stronger than natural fibers, PES fabric blends containing wool, cotton, rayon etc. are very popular because PES makes the fabrics more resilient and wrinkle free, while improving their shape-retention, draping characteristics and durability. However, dyeing of PES blends is a problem due to damage/reduction in the strength of natural fibers at high dyeing temperatures.[5-6] Similar to PES, low priced polypropylene (PP) is used a lot in technical textiles (filter, fleece, geo-textiles etc.), home textiles, the automobile industry, and to a lesser extent in apparel textiles etc. due to their special characteristics such as high heat resistance, low density, resistance to chemicals.[7-10] The dyeing of PP is very difficult and the fastness properties are not good due to its high crystalline molecular structure.

Although PES fabrics have many desirable bulk properties, they have some limitations on dyeing. It is well-known that polyesters are dyed with disperse dyes at high temperatures ($\approx$130°C) (sublimation becomes a more critical factor) in a closed system due to their hydrophobic nature.[11] The selection of dyestuffs is also limited to disperse dyes for polyesters because of their compact structure and high crystallinity.[12] The low and finite water solubility of these dyes is also a critical factor in determining leveling properties and the dyeing rate.[13] High temperature dyeing causes difficulties for polyester/natural blends because of the damage done to natural fibers during the dyeing process. Conventional polyester dyeing processes additionally require dispersing agents, dye carriers, and surfactants to disperse dye solubility in water. Due to dye reduction and migration, the fastness to washing was found to be on the satisfactory level dyeing with this dye.[14] Moreover, textile dyestuffs are rare, but nevertheless the most frequent textile allergy. A significant number of dyes (two thirds of all allergic dyes) potentially injurious to health belong to the class of disperse dyes.[15] The disperse dyes with an allergy-releasing effect are predominantly yellow, orange, red and blue dyes, which can also be hidden as mixtures in black textiles. Synthetic fibers can bind less firmly with certain types of dyes. This can in particular in damp environments lead to an increased migration of dyes from synthetic fiber fabrics possibly resulting in a contact allergy.[16]

In this context, there has been growing interest to find an alternative dyeing method (or alternative dyes) for PES and PP in recent years, as for example vat dyeing on PES and PP.[17] A lot of research activities have been developed in recent decades in order to improve the dyeability of such fabrics by the enhancement of hydrophobicity, cationization and gamma irradiation on PES for disperse dyes,[5] cationic surfactants on PES for disperse dyes,[18] the blending of polyamide 6 with PP,[19] the incorporation of hyperbranched polymers into the PP prior to fiber spinning.[10] Many authors have devoted their research to improving dyeability by incorporating functionalities on the fiber surfaces using environmentally friendly modifications such as plasma treatments.[17,20-23]

There has been very little attention focused on the application of hydrophilic acid dyes on hydrophobic PET fabrics. Milling acid dyes, which have excellent color brightness and very good wet fastness, can easily be applied to plasma-activated polyesters at a low temperature (≈80°C) within an hour of dyeing time where plasma modification is used as an alternative to the required pretreatment of PET textiles. Many authors have investigated the dyeability of plasma treated polyesters. Sarmadi et al. observed that the dyeability with basic dye (dyebath temperature 100°C and 2 hours dyeing time) can be improved by an increase in the time it is exposed to CF_4 cold plasma, and found *K/S* values between 0.50-1.51 for a 2% o.w.f. (on the weight of fabric) dark-shade dyeing with a hydrophobic basic dye.[24] Beside plasma activation by non-film forming gases, the applicability of plasma polymerization on textiles has recently been reviewed.[25] The dyeability with basic dye is enhanced on PET/cotton blends by *in situ* polymerization of acrylic acid and water as investigated by Öktem et al.[26] The work of Ferrero et al. has shown that the color fastness to washing with basic dye on PET by *in situ* polymerization of acrylic acid using low temperature plasma was found to be unsatisfactory probably because of an instable bond between grafted acrylic acid and dye molecules.[27] Anti-reflecting coating layers have been deposited with organo-silicon compounds using atmospheric plasma, which enhanced the color intensity on the PET surfaces as explained by Lee et al.[28] Okuno et al. studied the correlation between the crystallinity and dyeability of PET fibers using non-polymer forming gases by low temperature plasma.[29] They found that plasma treated samples significantly reduced the dyeability due to etching of macromolecules in the dyeable amorphous phase. Recently, Addamo reported that the color depth of air RF plasma treated PET fibers is related to their topographical characteristics and to their chemical surface composition.[30] They observed that the color strength (*K/S* value) with disperse dye at a dyeing temperature of 100°C can be increased by decreasing the fraction of light reflected from treated surfaces. Moreover, much research has also been done in the environmentally friendly dyeing of PET with disperse dyes in supercritical CO_2 which has the advantage of reducing the need for additional chemicals and waste water.[29,31-35] The scaling of the supercritical fluid dyeing experiment from laboratory size to industrial scale is far from a straight-forward procedure because it requires high pressure (260-300 bar). At low pressure the *K/S* value is decreased, which yields low dye solubility.[32] Moreover, conventional textile dyeing is dependent on the substrate materials due to their specific chemical nature. In a specific dyeing condition, dye molecules chemically and physically bind with the textile fibers and thus, dyestuff class is also limited to the chemical groups present in the fiber due to the dye-fiber interaction. For these reasons, substrate independent coloration is of particular interest not only for textiles, but also for the materials industries, where coloration is needed.

In this section, an attempt was made to solve some limitations of PES dyeing using hydrophilic acid dye by modifying the surface with plasma treatments. This approach, in particular, enables the removal of additional chemicals such as dispersing agents, dye carrier etc. during the PET dyeing. Surface modification of fabrics induced by NH_3/C_2H_2 was carried out in order to incorporate amine-end functional groups in the plasma film and consequently, provided accessible domains for the diffusion of hydrophilic acid dye molecules into the modified coating.

6.2. Experimental

6.2.1. Materials

Tightly woven and washed poly(ethylene terephthalate) (PET) fabric (76 ends/inch, 76 picks/inch, 43.5 g/m^2) used in this study was supplied by Sefar Inc., Switzerland. The gases Ar, O_2, C_2H_2, and NH_3 (purity 99.99 vol.-%) used for the plasma treatments were supplied by Carbagas, Switzerland. The acid dyeing agent used was C.I. Acid Blue 127:1 without further purification for the dyeing of plasma films.

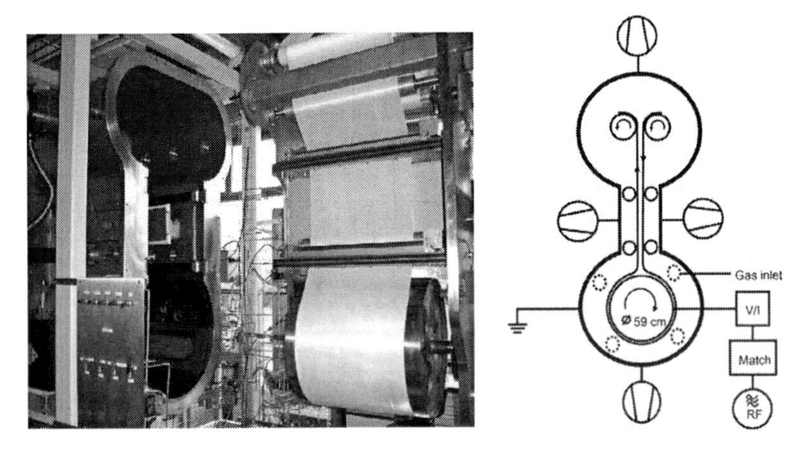

Figure 1. Set-up of web coater (left) and a schematic representation of the pilot-plant reactor (right).

6.2.2. Plasma Treatment

The semi-continuous web coater (reactor) used is shown in Figure 1 operating at 13.56 MHz (CESAR 1312, Dressler, Germany). The fabric samples (max. 65 cm width) were pulled continuously around the internal cylindrical electrode (drum of 59 cm in diameter) by take up rollers driven by a motor. Two RF electrodes were mounted in separate process chambers which enable a one-step process of e.g. pre-treatment and deposition. V/I probe measurements (ENI model 1640) indicated a power absorption of $\approx$70% within the plasma at rather symmetric conditions, i.e. low bias voltages. Different types of pumps were used: a rotary pump (Alcatel CIT, France) bringing the pressure down to 1 Pa and a turbo molecular pump (diffusion pump) (DCU 600, Pfeiffer Vacuum GmbH, Germany) reducing the pressure to 10^{-3} Pa. The working pressure and desired gas flow was monitored and adjusted by an adaptive pressure controller (VAT, Switzerland) and a multi gas controller 647B (MKS Instruments, Germany) respectively.

The chamber containing the fabric samples placed around the drum was cleaned and etched with Ar/O_2 (400 W, 10 Pa, 1 min) prior to the experiments. In order to obtain amine-incorporated functionalized coatings suited for acid dyeing, the plasma process parameters (power, gas flow, and exposure time) with a process gaseous mixture of NH_3/C_2H_2 were extensively investigated at a pressure of 10 Pa. The gases were mixed before introducing them into the chamber. A gas shower at four positions in the reaction chamber around the drum enabled a homogeneous treatment along the width and length of the fabrics. After the treatment, the samples were kept in a conditioned room (20$\pm$2°C, 65$\pm$2% RH) ready for dyeing and further experiments. The characterization of the polymerized film and film thickness on Si wafers was measured using an XPS analyzer (PHI LS 5600) and a surface profiler (HRP-75, KLA Tencor) respectively. The plasma process parameters and the various fastness properties of dyed PET are also discussed later in this chapter. The degree of hydrophilicity induced by the plasma treatment was measured on the basis of the static contact angle (CA) using G10 (Krüss GmbH, Germany).

6.2.3. XPS Analysis

XPS in combination with Ar sputtering provides additional information on the chemical composition of the plasma film including i) film growth mechanism on an atomic scale, ii) reactive functionalities within the film, iii) elemental surface composition etc. Surface sputtering allows depth profiling in order to determine the composition of a layer and to remove possible surface contaminations. However, it can influence the chemical composition by preferential sputtering.
In order to know the oxidation state in the C-N-H system, XPS analysis was carried out on coated Si wafers (100) at a base vacuum lower than 10^{-8} mbar with Mg Kα radiation (300 W, 1253.6 eV). XPS spectra were taken using multiple and survey scan modes with pass energies of 58.7, 187.8 eV, and step width using 0.025 and 0.80 eV/step, respectively. The in-depth compositional analysis was done by Ar^+ ion beam sputtering (3 kV, 20 mA). Typical sputtering rates on Ag were in the range of 2 nm/min. The incidence angle of the X-rays was 45°. Sputtering on a fabric raises different difficulties as the surface shows complex geometries and is not flat. Therefore, the sputtering of the coated fabrics would give a nonlinear response. In contrast, sputtering on a wafer reveals a direct correlation between sampling depth and elemental composition.

6.2.4. Contact Angle Measurement

Measurement of CAs is an easy way to characterize surface energy and surface hydrophilicity. Changes in surface hydrophilization in terms of CAs induced by plasma treatments were measured by the optical inspection method using G10 goniometer. Before the measurement all samples were conditioned at ambient temperature and the static CAs were measured with drop size of $\approx$10 μL with de-ionized water on coated glass substrates as well as on the coated textiles in a conditioned room (20°C and 65% RH). CAs on glass substrates enables the detection of small differences among the plasma treatments, whereas CAs on plasma-treated fabrics could not be measured, since the water drop was sucked in quickly (i.e. complete wetting). For each sample three measurements were taken at different locations and the average of these measurements were reported. An aging study of CAs was performed for a time period of several weeks.

6.2.5. Dyeing Procedure and Dyeing Mechanism

The dyeing of plasma treated samples was carried out in a laboratory-scale machine manufactured by Mathis, Switzerland (LABOMAT–8, Type BFA–8). Plasma treated fabrics were dyed with C.I. Acid Blue 127:1. The molecular structure of the dyestuff is shown in Figure 2. The dye molecules have a size approx. 3 nm. The light-shade dyeing was performed using 0.5% o.w.f. acid dye and 5% o.w.f. sodium sulphate salt for exhaustion and the pH of the dyebath was adjusted to 4.5–5 by adding ammonium sulphate. The liquor to fabric (L:R) ratio in dyeing was 1:50 and the following dyeing conditions were adopted: initial temperature 25°C, followed by temperature gradient of 1.5°C min^{-1} up to 80°C, then the dye bath temperature was maintained at 80°C for 60 min. This temperature required for the solubility and mobility of the dye molecules (see Chapter 7). The dyeing method is shown in Figure 3. However, the total dyeing time can be noticeably reduced, since the dye molecules do not have to penetrate into the polyester structure, but only into the nanoscaled coating.[36] After dyeing, the dyed fabrics were washed with soap (Ultravon W) at 60°C for 30 min (L:R=1:100), then rinsed with cold-hot-cold water and dried at room temperature. Thus, dye molecules that were not bound to the plasma coating were washed away.

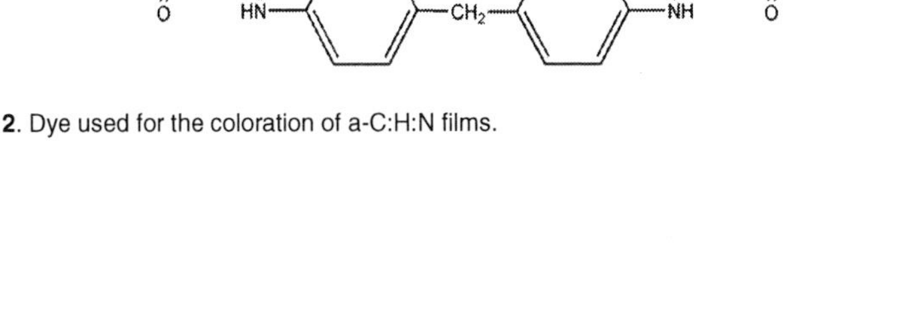

Figure 2. Dye used for the coloration of a-C:H:N films.

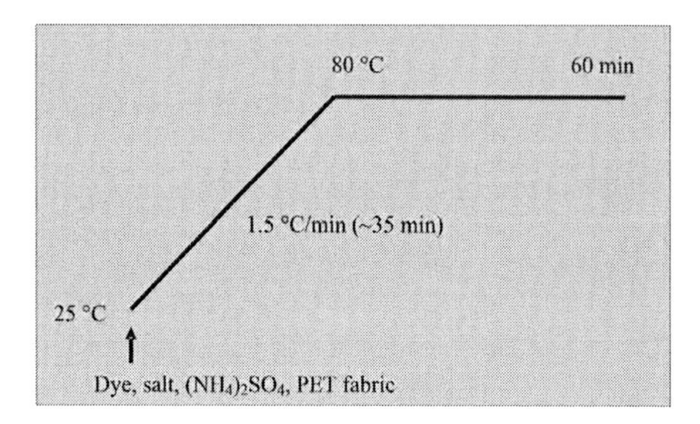

Figure 3. Dyeing procedure of plasma treated PET.

The substrate independent plasma dyeing mechanism is described in Figure 4.[37] The most probable interaction between film and dye is the ionic interaction of the dye ions with the amino groups of the films. Due to the presence of sulfonic acid groups, acid dye is water-soluble which is transported to the coating on the fiber by the motion of dye-liquor and/or the textile simultaneously in the exhaustion dyeing process. By an adsorption process dye molecules reach the coated surface; they are then diffused into the nanoporous structure of the coating. The basic amines within the plasma coating are decisive for acid dyeing which can be protonated in the acid medium and become fiber-cation (i.e. ammonium group etc.). Furthermore dye-molecules dissociate, which gives rise to dye anions, and as a consequence they interact with the cationic groups of the coated fiber mainly by the formation of ionic bonds. Secondary bonds such as dispersion, polar bonds and hydrogen bonds can also be formed between plasma polymer and dye.[11] As a result, dye molecules are fixed within the plasma coating, which is independent from the substrate materials.

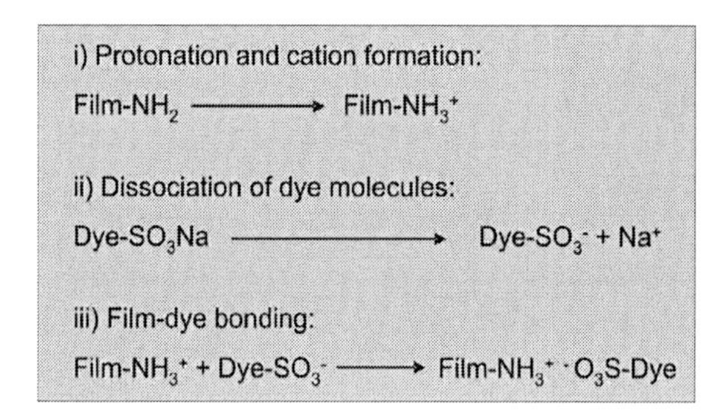

Figure 4. Film/dye interaction.[37]

6.2.6. Color Measurement

CIELAB color values (L^*, a^*, b^*, C^* and h^o) of the dyed fabrics were determined using a *Datacolor Spectraflash* interfaced to a PC. Each fabric was folded twice so as to give four thicknesses, and an average of six readings was taken for each measurement. The reflectance (*R*%) value of the dyed fabrics was measured over the wavelength range of 360-750 nm. The illuminant type was D65 and the observer angle was 10°. The color strength values (*K/S* values at 490 nm) of the fabrics were calculated from Kublka-Munk equation:[5]

$$K/S = (1 - R)^2/2R \tag{6.1}$$

where *K* is the absorption coefficient, *S* is the scattering coefficient, and *R* is the decimal fraction of the dyed fabrics.

6.3. Result and Discussion

6.3.1. Surface Modification of PET

An amorphous hydrogenated carbon film (a-C:H hard coatings) can be obtained using the polymerizable gas C_2H_2; while the structural modification of a-C:H films by the addition of nitrogen to the hydrocarbon precursor yielded hydrophilic functional sites (mainly amine functionalities) in a-C:H:N coatings.[36] In general, the a-C:H:N films become more graphitic and the density of voids and/or porosity increases with the incorporation of nitrogen and/or nitrogen functionalities in the coating.[38,39] Moreover, the gaseous mixture assures a textured surface rather than a smooth one. In the previous chapter, the mass deposition rates for a wide range of NH_3/C_2H_2 ratio was examined.[40] It was found that plasma deposition, in particular radical-promoted plasma, was governed by the composite parameter: power input per monomer flow *W/F*. At moderate energy input ($2 \leq W/F \leq 2.4$ W/sccm) a maximum deposition rate can be achieved, as can be seen in Figure 5. In general, increasing the C_2H_2 content in the gas mixture leads to an increase of hydrocarbon radicals in the active plasma zone resulting in a gradual increase in a-C:H film character and enhanced deposition rates.[40] Modifications in film growth such as degradation reactions of polymer chains, chemical and physical etching, and some temperature effects can be observed at higher specific energies ($W/F > 2.4$ W/sccm) yielding a reduced deposition rate due to the transition between film growth and erosion regime.[25,36,40] In addition, at very low flow rates the film growth is limited by the availability of monomer supply.[41] Commonly, in low pressure plasma the energetic particles can be accelerated much more and long-living radicals are generated compared to atmospheric pressure plasma. As a consequence etching effects can play a major role in low pressure plasma. However, the application of ammonia/acetylene plasma is very beneficial in avoiding strong etching effects during plasma polymerization, while coherent etching yield the formation of voids within the growing films.[42] It can be concluded that the optimum chemical modification can be attained at moderate energy input, and suitable C_2H_2/NH_3 ratio (≈ 1.0) to obtain accessible amine-functional groups within a-C:H:N coatings.[43]

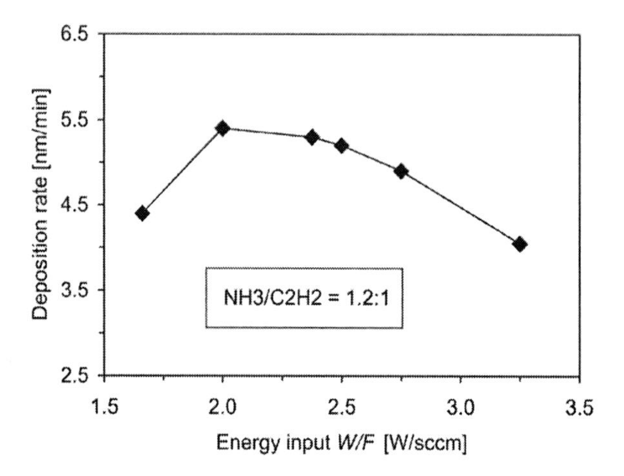

Figure 5. Effect of energy input *W/F* (RF power/C_2H_2 flow) on deposited plasma film.

6.3.2. XPS Characterization

The XPS studies provided further insight into the chemical nature of the surface of the plasma coating. In order to observe nitrogen incorporation inside the film, sputtering at five different time frames namely 0, 15, 30, 45 and 60s was examined and C1s, N1s and O1s peaks were found (Figures 6a to 6c). It can be seen from C1s peak that the carbon network (C-C, C-H, C=N, C-N, C=O, and COO etc.) increases with the increase in sputtering time.[43] In contrast, the opposite phenomena was observed for the N1s peak, with the nitrogen incorporation (N-C, N=C, and N-O etc.) decreasing as the sputtering time increased (from 22 to 10 at%). Since the deposition conditions were kept constant and uniform deposition conditions are assumed due to homogeneous dyeing results (see Figure 8), preferential sputtering effects of weaker nitrogen groups during XPS measurements within a crosslinked hydrocarbon matrix can be assumed. In general, nitrogen is a replacement of carbon in the plasma film yielding nano-structured (porous) crosslinked surfaces. This analysis is in good agreement with results of other groups.[39,41]. Oxidized surfaces are probably due to incorporation of C=O, COO, and N-O etc. groups during post-plasma reactions in contact with the atmosphere prior to the XPS analysis.[43] As these hydrophilic groups stay on the topmost layer on the surface, after a short sputtering time the oxygen functionalities detected were below 2 at%, as shown in Figure 6c indicating less oxidation within the nanoporous structure and thus less radical sites left after plasma polymerization.

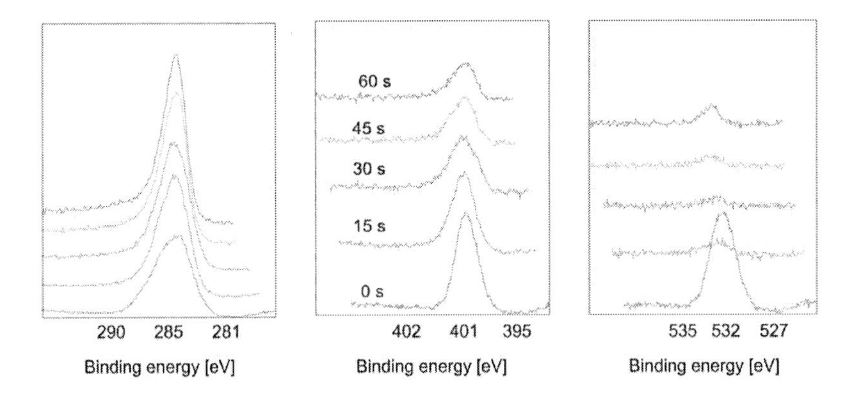

Figure 6. C1s (left), N1s (middle) and O1s (right) spectra in a-C:H:N film for different sputtering times: 0, 15, 30, 45, and 60s (*W/F* = ≈2.4 W/sccm, NH_3/C_2H_2 = 1.2:1).

6.3.3. Contact Angle

The untreated PET showed a hydrophobic nature with a CA of ≈80° due to the textured surface and some manufacturing residuals. Wettability of PET fabrics was improved markedly: the water drop on the surface disappeared immediately in all cases, so that the static CA could not be given as approximately 0° (complete wetting). To obtain an effective CA on textile structures, a more sophisticated analysis can be performed.[4,44] In order to analyze the surface hydrophilic effect on the different treatment conditions, the static CAs on glass substrates were also measured. Figure 7 shows the CA as a function of aging for different gas ratios. It is evident that at gas ratio>2.7 there is a large decrease in CA below 20° even after 30 days aging. As discussed previously, the optimum plasma deposition was obtained at moderate energy input *W/F* ≈2.4 and a gas ratio of around 1.0, to obtain stable nanoporous a-C:H:N coatings which is supported by the observation of CA data (around 50°) and showing higher *K/S* values. This result indicated that the incorporation of hydrophilic polar groups was decreased at the topmost surface layer at these conditions. The CAs increased, but were still below 60°, with the decrease in gas ratio (Figure 7) due to increased monomer fragmentation accompanied by an increase in deposition rate. After two weeks of storage time the CAs turned into almost constant for these coatings as the surface reorientation and reorganization stabilized (lower number of polar functionalities at the film surface). On the other hand, at gas ratio>2.7 the high number of polar functionalities resulted in a slow gradual increase in CA with storage times up to 30 days due to restructuring on the modified amorphous surfaces or simply hydrocarbon adsorption from the environment. Several groups, on the other hand, have reported stronger and faster reorganization observations for PET films and PET fibers by non-polymerizable RF plasma (e.g. Ar/O_2) (see Chapter 3 and 4).[4,45,46] In these conditions, the dyeability was found to be low due to less number of amines incorporation within the coatings. Thus, coloration does not depend on the surface wettability, but on the nanoporous structure.

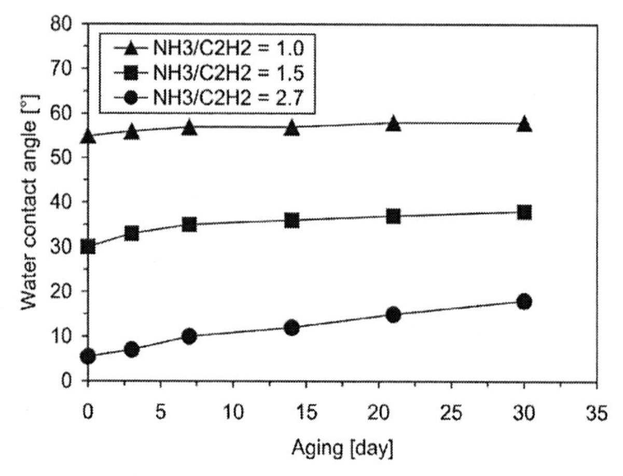

Figure 7. Water contact angle (static) on plasma treated glass substrates with aging for different energy inputs *W/F* (RF power/C_2H_2 flow).

6.3.4. Surface Dyeability

In general, hydrophilic acid dye can be applied to hydrophobic PET in the presence of hydrophilic functional/reactive sites such as amines etc. in a modified coating induced by plasma treatments. Plasma polymerization significantly influences surface charges on the substrate due to the addition of reactive functionalities on the modified surfaces. The positively charged amine end-groups are incorporated in the plasma deposited film; therefore anionic dye was used to dye the film. It is evident that the dyeability is independent from the substrate material as the plasma surface film is dyed only without changing the properties of the bulk textile. The dissolved dyes adsorbed onto the thin film then easily diffuse, as these dye molecules are small in size, into the water-filled pores according to the "pore model" instead of the "free volume model".[47] They are able to form an ionic bond with the amine end-groups. In this case, the dyeing principles are similar to those in natural fiber dyeing. In conventional dyeing, the "free volume model" applies to hydrophobic synthetic fibers and the dye molecules diffuse above the glass transition temperature (T_g) through the holes in amorphous region of the synthetics (such as polyester).

Dyeability is improved markedly with a relative color strength value up to 1.2 for a 0.5% o.w.f. light-shade dyeing. This value strongly depends on the plasma deposited film, whereas the *K/S* value was zero for untreated PET, as no dyeing was possible. As shown in Figure 8, the dye uptake increased almost linearly with the increase in film thickness.[28] The gaseous mixture helped to build up voids in the deposition resulting in a porous, nano-structured-graphitic film during the plasma polymerization, which were accessible to dye molecules throughout the entire film volume.[39,48] A level dyeing was obtained because of the homogenous distribution of polar groups everywhere in the film, thus proving consistent uniform plasma deposition conditions.

The relative color strength value was increased gradually with the increase in plasma exposure time, as can be seen in Figure 8. It is noteworthy that with increasing plasma process time the penetration of reactive plasma species yielding hydrophilization effects by plasma polymerization goes deeper into the textile structure, even in the inter-filament or inter-yarn spaces, resulting in better dyeability. The deposition rate is directly proportional to the treatment time. On the other hand, the reduced film thickness at very low flow rates led to a lower *K/S* value of the PET surfaces. This result is consistent with the report of Okuno et al.[29] It is evident that the coating obtained was very stable and had very good adhesion with the fabric surface, as was observed by using the Abrasion & Pilling Tester (NU-Martindale, James H. Heal, England). No damage to the film and fibers on the surface of the plasma treated and dyed PET was detected after 60,000 cycles (Test method, SN 198514). Recently (see Chapter 8), a reduced yellowness index was found using ammonia/ethylene mixtures instead.[37]

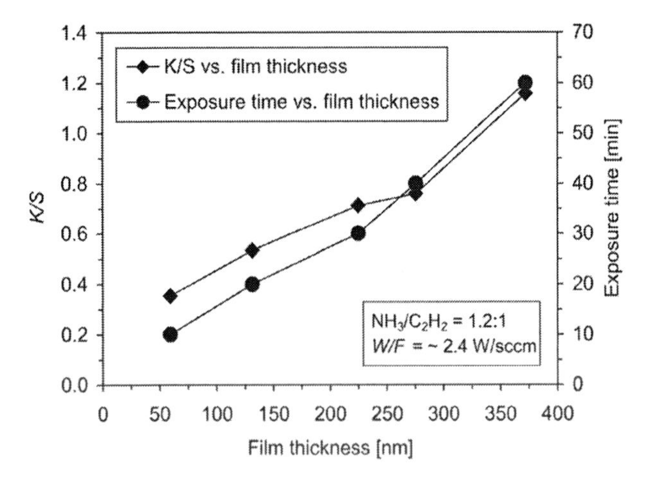

Figure 8. Relative color strength values (*K/S*) depending on film thickness and exposure time (light-shade dyeing with 0.5% o.w.f. acid dyes).

The change in color was also investigated on the basis of CIELAB color spaces in terms of colorimetric data (L^*, a^*, b^*, C^* and h^o) at different exposure times and with the same energy input at constant pressure. Table 1 shows that the change in L^* values were closely related to the change in h^o values. While a decrease of the L^* and h^o values of the dyed samples correlates with increasing exposure time, no significant change in C^* was observed. A similar trend was reported for aramid fabrics with C.I. Disperse Blue 56 by sputter-etching.[49] While there was little change for the red/green component a^*, the blue/yellow component b^* increased rapidly in the color with each increase in exposure time. This implies that the yellowness index increases with time due to more unsaturated bonds on the surface,[43] which is one disadvantage of using acetylene (HC≡CH) as film-forming monomer.

In conclusion, the color depth of plasma treated PET fabrics can be improved markedly depending on the film thickness and the amine functionalities inside the deposited film. Moreover, the depth of shade of the dyed fabrics is influenced by the surface morphology and in contrast, dark shade is difficult to obtain due to the smooth surface of synthetic fibers. The plasma modification enhances the surface roughness and adhesion of PET leading to a lower light reflection (*R%*) due to interference and a scattering effect, as a consequence of the higher *K/S* value.[49-51]

Table 1. Effect of plasma exposure time on light-shade dyed PET with 0.5% o.w.f. acid dyes.

Plasma parameters					Color coordinators (D65/10)				
Power (W)	Pressure (Pa)	Time (min)	C_2H_2 (sccm)	NH_3 (sccm)	L^*	a^*	b^*	C^*	h°
475	10	10	200	250	65.95	-11.08	-5.45	12.35	206.21
475	10	20	200	250	61.78	-11.79	-2.67	12.09	192.75
475	10	30	200	250	60.08	-10.62	2.12	10.83	168.69
475	10	40	200	250	57.20	-11.95	3.10	12.00	185.27
475	10	60	200	250	55.38	-9.02	7.57	11.78	140.00

6.3.5. Fastness Properties

The washing color fastness of the dyed samples was assessed using the ISO test method. Under D65 illumination color changes, staining and rub were evaluated using grey scales: ISO-105-A02 grey scale for assessing change in color; ISO-105-A03 grey scale for staining and rub. Table 2 shows that the acid dyeing on plasma treated samples displayed acceptable fastness properties to laundering and rubbing. It is clear that no significant difference in both fastness properties were detected. This result demonstrates that a permanent ionic bond was obtained between dye molecules and amine-end groups in the functionalized coating.

Table 2. Color fastness properties of plasma-treated dyed PET fabrics (W/F = $\approx$2.4 W/sccm, NH_3/C_2H_2 = 1.2:1).

Exposure time (min)	Color change	Staining		Rub	
		Cotton	Wool	Dry	Wet
10	3-4	4	3-4	4	3
20	3	4	3	4	3
30	3-4	3-4	3-4	4	3
40	3	3-4	3	4	3
60	4	4	4	4-5	3

6.4. Conclusion

A significant surface modification of PET fabrics was obtained using low pressure RF plasma polymerization. The gaseous mixture of NH_3/C_2H_2 assured a homogenous functionalized coating containing hydrophilic amine-end groups on PET surfaces, and nitrogen incorporation in the a-C:H film indicated a porous structured a-C:H:N coating composed of highly crosslinked and branched surface (hydrocarbon chains) due to the replacement of pure carbon network.[52] The film growth was strongly influenced by the energy input *W/F*. A moderate energy input (*W/F* between 2.0 and

2.4 W/sccm for the used web coater) gave the optimum film deposition with respect to dye uptake and thus functional group density.

Plasma modification is a promising technique to change the surface properties of PET from hydrophobic to hydrophilic and to enable the dyeing of PET with hydrophilic acid dyestuff at low temperatures. The plasma treatment significantly improved the dyeability and color fastness properties of dyed PET. As the a-C:H:N films became more graphitic and porous, the dye molecules diffused into the hydrophilic nanoporous film and formed an ionic bond with the amine-end groups. The dye uptake was strongly correlated with plasma process time, in other words with the deposited film thickness. It is interesting to find that the *K/S* values can be controlled by controlling film thickness during plasma polymerization. Moreover, the energy input and gas ratio, which have a strong influence on film growth and film structure, should be considered to obtain higher *K/S* values. Since color strength values are enhanced with the increased CA, hydrophilicity (of the outermost surface) does not influence on surface (plasma film) dyeability. It is obvious, since merely the functional film is dyed, the entire dyeing process is independent from the substrate materials. Hence, the same dyeing principle can also be applied to all hydrophobic synthetic textiles as it has already been proved for PP, acetate, aramid etc. This study has explored a new method to dye hydrophobic polyesters in an environmentally friendly way. This ecological process reduces the need for additional chemicals, waste water etc. and a corresponding reduction in the cost of effluent treatment. Having diversified dyestuff molecules small in size will be useful for dyeing plasma modified film and will facilitate the commercial application of these findings in industry. In addition, the dyeability of the nanoporous coatings with dye molecules of around 3 nm size proves the open structure of the nano-pores. Furthermore, ammonia/ethylene plasma can also be used to deposit nanoporous ultra thin films with high amine functionalities, thus the dyeing time can be shortened greatly, as described in the following chapter.

6.5. References

[1] M. M. Hossain, J. Müssig, A. S. Herrmann, D. Hegemann, *J. Appl. Polym. Sci.* **2009**, *111*, 2545.
[2] N. Inagaki, K. Narushim, N. Tuchida, K. Miyazaki, *J. Polym. Sci., Part B: Polym. Phys.* **2004**, *42*, 3727.
[3] Y. Cui, N. Yoon, *Dyes Pigments* **2003**, *58*, 121.
[4] M. M. Hossain, D. Hegemann, P. Chabrecek, A. S. Herrmann, *J. Appl. Polym. Sci.* **2006**, *102*, 1452.
[5] M. H. Zohdy, *Radiat. Phys. Chem.* **2005**, *73*, 101.
[6] V. A. Shenai, M. C. Sadhu, *J. Appl. Polym. Sci.* **1976**, *16 (11)*, 2057.
[7] W. Oppermann, H. Herlinger, D. Fiebig, O. Staudenmayer, *Textilveredlung* **1996**, *9*, 588.
[8] H. Herlinger, G. Augenstein, U. Einsele, *Melliand Textilberichte* **1992**, *9*, 737.
[9] M. Muskatell, L. Utevski, M. Shenker, S. Daren, M. Peled, Y. Charit, *J. Appl. Polym. Sci.* **1997**, *64*, 601.
[10] S. M. Burkinshaw, P.E. Froehling, M. Mignanelli, *Dyes Pigments* **2002**, *53*, 229.
[11] Industrial Dyes: Chemistry, Properties, Applications, ed. Hunger, K., Wiley-VCH, Weinheim, **2003**, pp. 392-408.
[12] M. R. De Giorgi, E. Cadoni, D. Maricca, A. Piras, *Dyes Pigments* **2000**, *45*, 75.
[13] S. V. Kulkarni, C. D. Blackwell, A. L. Blackard, C. W. Stackhouse, M. W. Alexander, Chemistry, equipment, procedures, and environmental aspects, Radian corporation, Research triangle park, North Carolina, Noyes Publications, USA, **1986**, pp. 61-64.
[14] Y. A. Son, J. P. Hong, T. K. Kim, *Dyes Pigments* **2004**, *61*, 263.
[15] M. Kalcklösch, H. Wohlgemuth, M. Kunze, BIFAU, Umweltreihe Heft 15, Textilallergie, Berliner Institut für Analytik und Umweltforschung e.V. Berlin, **1999**, pp. 88-91.
[16] F. Klaschka, *Melliand Textilberichte*, **1994**, *75 (3)*, 196.
[17] F. Gähr, *Technische Textilien* **2006**, *2*, 98.
[18] T. Choi, Y. Shimizu, H. Shirai, K. Hamada, *Dyes Pigments* **2001**, *50*, 55.
[19] A. Seves, G. Testa, B. Marcandalli, L. Bergamasco, G. Munaretto, P. L. Beltrame, *Dyes Pigments* **1997**, *35 (4)*, 367.
[20] G. S. Kakad, A. R. Rathod, B. Suman, *Man-made Textiles India* **2006**, *3*, 85.
[21] C. J. Jahagirdar, L. B. Tiwari, *J. Appl. Polym. Sci.* **2004**, *94*, 2014.
[22] C. Riccardi, R. Barni, E. Selli, G. Mazzone, M. R. Massacra, B. Marcandalli, G. Poletti, *Appl. Surf. Sci.* **2003**, *211*, 386.
[23] U. Vohrer, M. Müller, C. Oehr, *Surf. Coat. Technol.* **1998**, *98*, 1128.
[24] M. Sarmadi, Y. AH. Kwon, *Modifying Fiber Properties* **1993**, *25*, 33.
[25] D. Hegemann, *Indian J. Fibre Text. Res.* **2006**, *31*, 99.
[26] T. Öktem, N. Seventekin, H. Ayhan, E. Piskin, *Indian J. Fibre Text.* **2002**, *27*, 161.
[27] F. Ferrero, C. Tonin, R. Peila, F. R. Pollone, *Color. Technol.* **2004**, *120*, 31.
[28] H. R. Lee, D. J. Kim, K. H. Lee, *Surf. Coat. Technol.* **2001**, *142-144*, 468.
[29] T. Okuno, T. Yasuda, H. Yasuda, *Text. Res. J.* **1992**, *62 (8)*, 474.
[30] A. Raffaele-Addamo, E. Selli, R. Barni, C. Riccardi, F. Orsini, G. Poletti, L. Meda, M. R. Massafra, B. Marcandalli, *Appl. Surf. Sci.* **2006**, *252*, 2265.
[31] G. Montero, D. Hinks, J. Hooker, *J. Supercrit. Fluid.* **2003**, *26*, 47.

[32] A. S. Özcan, A. Özcan, *J. Supercriti. Fluid.* **2005**, *35*, 133.
[33] M. Banchero, A. Ferri, *J. Supercrit. Fluid.* **2005**, *35*, 157.
[34] A. Hou, K. Xie, J. Dai, *J. Appl. Polym. Sci.* **2004**, *92*, 2008.
[35] O. S. Fleming, S. G. Kazarian, E. Bach, E. Schollmeyer, *Polymer* **2005**, *46*, 2943.
[36] M. M. Hossain, A. S. Herrmann, D. Hegemann, *Plasma Process. Polym.* **2007**, *4*, S1068.
[37] D. Jocic, S. Vilchez, T. Topalovic, A. Navarro, P. Jovancic, M. R. Julia, P. Erra, *Carbohyd. Polym.* **2005,** *60*, 51.
[38] N. K. Cuong, M. Tahara, N. Yamauchi, T. Sone, *Surf. Coat. Technol.* **2005**, *193*, 283.
[39] F. L. Freire, G. Mariotto, R. S. Brusa, A. Zecca, C. A. Achete, *Diam. Relat. Mater.* **1995**, *4*, 499.
[40] D. Hegemann, M. M. Hossain, *Plasma Process. Polym.* **2005**, *2*, 554.
[41] D. A. Waldman, Y. L. Zou, A. N. Netravali, *J. Adhes. Sci. Technol.* **1995**, *9 (11)*, 1475.
[42] B. Z. Jang, *Compos. Sci. Technol.* **1992**, *44*, 333.
[43] M. M. Hossain, A. S. Herrmann, D. Hegemann, *Plasma Process. Polym.* **2007**, *4*, 135.
[44] M. M. Hossain, A. S. Herrmann, D. Hegemann, *Plasma Process. Polym.* **2006**, *3*, 299.
[45] S. Carlotti, A. Mas, *J. Appl. Polym. Sci.* **1998**, *69*, 2321.
[46] B. Gupta, J. Hilborn, CH. Hollenstein, C. J. G. Plummer, R. Houriet, N. Xanthopoulos, *J. Appl. Polym. Sci.* **2000**, *78*, 1083.
[47] "*Chemical Principles of Synthetic Fibres Dyeing*", S. M. Burkinshaw, Eds., Blackie, Glasgow **1995**.
[48] P. Hammer, F. Alvarez, *Thin Solid Films* **2001**, *398-399*, 116.
[49] T. Wakida, S. Tokino, *Indian J. Fibre Text.* **1996**, *21*, 69.
[50] J. Jang, Y. Jeong, *Dyes Pigments* **2006**, *69*, 137.
[51] D. Knittel, E. Schollmeyer, *Polym. Int.* **1998**, *45 (1)*, 110.
[52] Y. L. Zou, A. N. Netravali, *J. Adhes. Sci. Technol.* **1995**, *9 (11)*, 1505.

7. Deposition of Nanoporous Ultrathin Films Using NH_3/C_2H_4 Discharges

7.1. Introduction

Plasma polymers from NH_3/C_2H_2 gaseous mixtures enabled PET dyeing with hydrophilic acid dye, as demonstrated in Chapter 6. In this section, the novel approach is explored to develop a new process to deposit amine-embedded coatings using ammonia/ethylene (NH_3/C_2H_4) gaseous mixtures for the substrate independent surface dyeing.[1]

Nanoscaled composite thin layers were deposited on PES textiles by low temperature RF plasma. Since the coating approach is largely independent from the substrate materials, it has an enormous potential in industrial applications. In this process, amine terminating groups, NH_2 and eventually NH, were embedded in hydrogenated carbon films (a-C:H films). Incorporation of nitrogen functionalities in the coatings provided crosslinked and a porous structure which facilitated dye molecule access to amines inside the coatings, thus dyeing time can be shortened noticeably. The surface morphology of the ultrathin layers was characterized by Atomic Force Microscopy (AFM) and BET (Brunauer, Emmett and Teller method) measurements. The chemical composition of the plasma films was characterized by XPS. The deposited films were dyed with water soluble acid dyes. The dyeing parameters, such as dyeing temperature, dyeing time, and pH, were systematically optimized in order to obtain higher color yield, level dyeing, and higher fastness properties. The color intensity is described in terms of the CIELAB color spaces which were obtained with a spectrophotometer.

7.2. Experimental

7.2.1. Materials

The PES fabric, dyes and gases used in this study were described in Chapter 6. The C_2H_4 gas (purity 99.99 vol.-%) used for the plasma treatments were also supplied by Carbagas, Switzerland.

7.2.2. Material Characterization

The characteristics of the fabric surface were examined by SEM (Hitachi, S-4800) and XPS.

7.2.3. Deposition of Plasma Film

A RF plasma generator was used for the deposition of nitrogen incorporated a-C:H films on PES textiles using NH_3/C_2H_4 gaseous mixtures. In order to obtain nanoporous thin films, different modifications were employed by varying the

important plasma parameters such as discharge power and gas flow ratios. The plasma treatments were carried out in a pilot plant reactor to demonstrate the feasibility for industrial scaling-up. The reactor is described in more detail in the previous chapter.[2]
The fabric samples were kept on the cylindrical electrode (65 cm width). The RF power was connected to the electrode and the glow discharges were carried out for the required power, gas flow and duration while the pressure was kept at 0.1 mbar for all experiments. Prior to the deposition of a-C:H:N films, the fabric samples were cleaned and activated with Ar/O_2 (1 min, 400 W) plasma in order to enhance adhesion between plasma polymers and substrate surfaces.

7.2.4. AFM Imaging

The characterization of the deposited films (thickness in the nanometer range) on the fabrics is rather difficult due to their complex geometries compared to solid polymer surfaces. Therefore, the topographical changes and the surface chemistry of the coatings were measured on Si wafers (100) in order to compare with the textile substrates. The topographical changes of plasma-coated surfaces were characterized by AFM, in non-contact mode using a very fine tip. Film thickness was determined on a film edge by measuring the step-height.

7.2.5. XPS Analysis

XPS analysis of the C-N-H system was performed according to the conditions described in Chapter 6.

7.2.6. BET Measurement

The specific surface area was obtained based on physical adsorption of nitrogen on the nanoporous coated PES surface, using the BET method. The monolayer surface adsorption was determined, and then by assuming a surface area occupied by an adsorbed molecule it became possible to determine the coated surface area. The nitrogen adsorption isotherms of coated samples were obtained at 77 K using a Beckman Coulter Adsorption Analyzer (SA 3100). The samples were degassed by heating at 105°C for 24 h prior to the N_2 adsorption measurement. Adsorption data of N_2 in the relative pressure region (P/P_O) from 0.025 to 0.082 were used to fit the BET equation. An uncoated PES fabric was measured as a control.

7.2.7. Plasma-Films Dyeing

The light-shade dyeing was performed according to the conditions described in Chapter 6. At the temperature between 20 to 120°C the dyeing was maintained for 1, 5, 10, 20, 60, 90, and 120 min; and a pH in the range of 4.5 to 7.0.

7.2.8. Color Measurement

The color intensity of the dyed fabrics was determined according to the methods described in Chapter 6.

7.3. Result and Discussion

7.3.1. Nanoporous Functionalized Films

The nitrogen incorporation and etching of a-C:H films yielded nanoporous ultrathin coatings (<80 nm) which depended mainly on gas ratio and plasma power. The admixtures of reactive ammonia into the monomer gas caused strong structural disorder at higher NH_3/C_2H_4 ratios.[3] As a consequence, a reduced porosity yielded low dye uptake. Moderating energy input and lowering NH_3/C_2H_4 ratios (around 1:1), on the other hand, assured plasma polymerization in the regime where both deposition and etching took place. Thus, a nanoporous and crosslinked network with accessible functional groups was obtained. The surface roughness increased slightly as compared to Si wafers, but remained 1.0 nm due to some ion bombardment effects. The peak to peak roughness was enhanced to 1.5 nm probably because of chemical etching leading to surface texturing. As shown in Figure 1, the AFM images indicate that the interconnected voids in the coating are below 25 nm. The dye molecules are thus small enough (about a few nanometer) to diffuse easily through the interconnected voids of the nanoporous structure into the plasma-polymer matrix and form dye-film bonds.

Higher energy input avoids the formation of voids, with high crosslinking of the amorphous network, and raises film rigidity due to the hybridization state of carbon atoms in the a-C:H:N system. The presence of clustered carbon atoms decreases in the amorphous network connectivity. As a consequence, at high energy input more crosslinked and dense coating cause declination of dye molecules penetration into the coating during plasma dyeing process (see Figure 4).

Figure 1. Surface topography of a-C:H:N plasma coating (600 W, NH_3/C_2H_4 = 0.84, 20 min): left, surface topography, and right, 3-D surface image.

To estimate the surface area of a nanoporous coating on PES textiles, a BET adsorption isotherm was obtained from nitrogen adsorption, as shown in Figure 2. The adsorption behavior in the low *P/Po* (<0.09) region was attributed to monolayer-multilayer adsorption. The monolayer coverage occurred between *P/Po* = 0.025 and 0.082. A plot of BET analysis (not shown in Figure) from Figure 2 yielded a linear portion, whose slope and intercepts gave the monolayer capacity of the sample and BET constant. The total surface of the sample calculated from these findings was 0.1 m^2/g corresponding to the mass of the fabric, where the BET surface of the uncoated PES fabric was noticeably smaller (Figure 2). Considering the mass of the nanoporous coating, a BET surface of minimum 30 m^2/g was obtained due to the high film porosity. The high surface thus enables a high number of functionalities embedded in the coating.

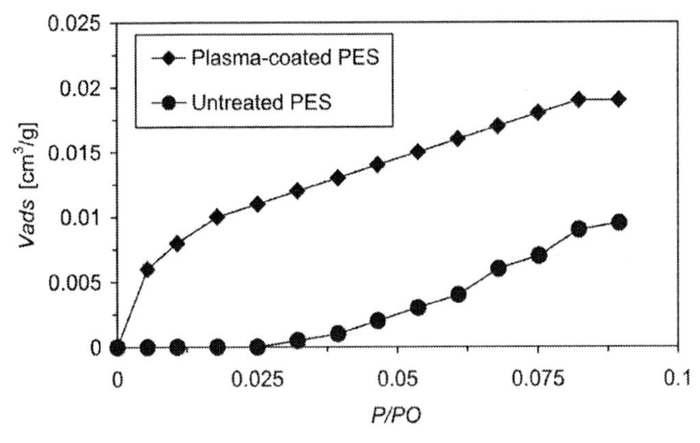

Figure 2. Adsorption isotherm for nitrogen at 77 K of a-C:H:N films (600 W, NH_3/C_2H_4 = 0.84, 20 min), where P/P_O is relative pressure and V_{ads}, adsorbed volume of N_2 molecules.

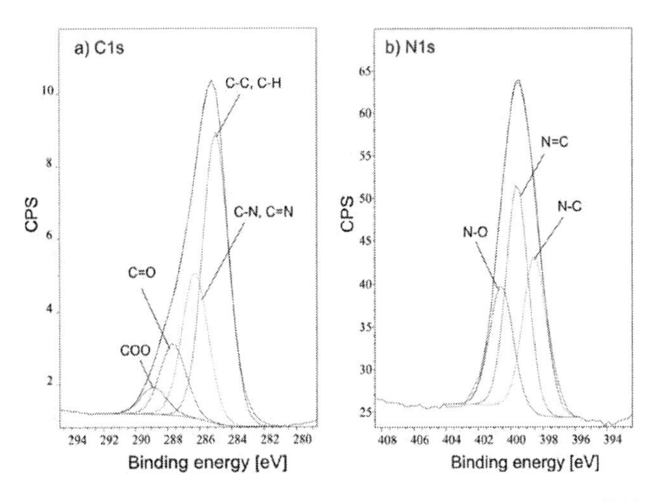

Figure 3. Deconvoluted C1s and N1s XPS spectra of a-C:H:N films (600 W, NH_3/C_2H_4 = 0.84, 20 min).

7.3.2. XPS Characterization

The C1s and N1s XPS spectra for the coated Si wafer are shown in Figure 3a-3b. The C1s signal can be fitted with four assignments. It shows that the intensity of the first peak corresponds to C-C or C-H network at 285 eV. The peak at 286.3 eV is assigned to the C-N or C=N groups,[4] which are most likely from incorporation of C-NH_x functionalities such as primary and secondary amines.[5] The other two assignments, having binding energies of about 287.6 and 288.9 eV correspond to C=O, and COO bonded network, respectively. Similarly, the N1s peaks were also deconvoluted into three assignments corresponding to N-C, N=C, and N-O bonds, which are situated at about 398.5, 399.6, and 400.6 eV (Figure 3b).[5,6] Since free radicals are created during plasma polymerization, oxygen containing assignments such as C=O, N-O and COO are probably formed on the coated surface during post-plasma reaction at atmosphere before XPS analysis.

7.3.3. Surface Dyeability

The coloration was found to increase with the increase in plasma exposure time, i.e. dyeability strongly depends on film thickness. Thus, higher color yield indicated a higher number of accessible nitrogen functional groups within the nanoporous hydrocarbon matrix. A linear relationship proved that coloration was achieved throughout the entire film volume.[7] Nitrogen functionalities, mainly terminating amine groups within the plasma polymer,[8-10] facilitated the superficial dyeing of the PES using acid dyestuffs. Adsorption of acid dyes was attributed to electrostatic attraction between amine groups and dye anions, and van der Waals forces. Dye molecules diffused and penetrated into the pores of the films and chemically bonded with the amine functional groups. Thus, the plasma coating was dyed throughout the entire film thickness with this dye class, whereas the whole dyeing process was independent from the PES textiles. Coloration of PES with this dye is generally impossible, since untreated PES does not contain amines in their structure, which would be needed for a dye-fiber reaction. Figure 4 demonstrates the color intensity given by K/S values normalized to the plasma exposure time of the treated and dyed PES as a function of energy input W/F (energy per monomer flow) and gas ratio. It illustrates that color intensity is correlated both to gas ratio and energy input. Higher ammonia to hydrocarbon ratios led to more etching during plasma polymerization, as indicated by the reduced deposition rate (see Chapter 8).[11] These conditions might have reduced the density of accessible amine functionalities within the coatings by higher fragmentation and crosslinking, since color strength per plasma exposure time was found to be very small (<0.003/nm). A low ammonia flow rate in the monomer gas, on the other hand, yielded low coloration due to the insufficient number of nitrogen functional groups in the plasma polymer. Therefore, the ammonia flow in the monomer gas should be in the range where both incorporation of nitrogen functionalities and formation of a nanoporous structure can take place.

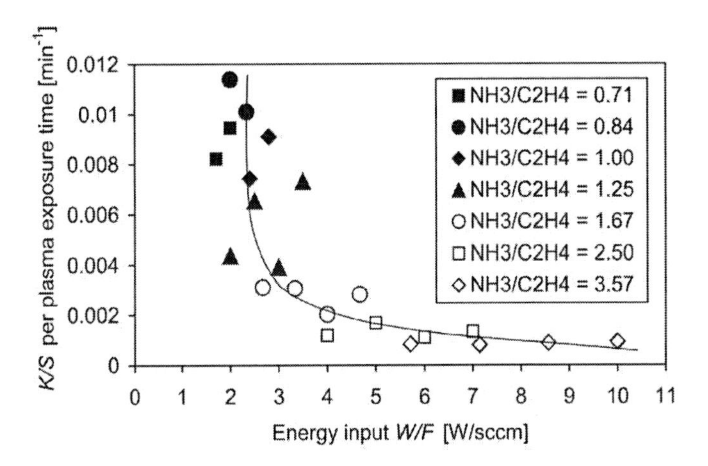

Figure 4. Color intensity (*K/S*) of plasma-treated and light-shade (0.5% o.w.f.) dyed PES textiles depending on energy input *W/F* (normalized to plasma exposure time).

At rather low energy input *W/F* a reduced dyeability was observed caused by low molecular fragmentation. At moderate *W/F* in the range of 2-3 W/sccm (120-180 J/cm^3) excellent *K/S* values were obtained, as can be seen in Figure 4. The results proved that this energy range assured plasma polymerization in a regime where both deposition and etching processes took place, thus yielding a nanoporous and crosslinked network with a maximum incorporation of reactive nitrogen functionalities inside the coating. Furthermore, dyeability was found to decrease again at around *W/F*>3 W/sccm (>180 J/cm^3) indicative of strong fragmentation yielding relatively hard coatings (less voids) and less accessible nitrogen functionalities in the coating showing low dye uptake.

7.3.4. Rate of Dye Uptake (*K/S*)

Uptake of acid dyes onto the plasma coated fabric enhanced remarkably even in a very short dyeing time (5 min), as shown in Figure 5a. It is very interesting to see that the dye uptake remained similar from a short dyeing time (5 min) to a very long dyeing time (120 min), since no significant difference in dye uptake was found at longer time periods. The observed enhancement of dye uptake can be attributed to the deposition of nanoporous thin films which provide amine groups easily accessible to dye molecules within a short time. This demonstrates that the dye diffusion coefficients are quite high; in general these values are quite low when traditional dyeing processes are used.[12] In addition, since ultrathin films were dyed, the dye uptake reached to “dyeing equilibrium” very fast. Moreover, the uniform shade obtained certainly proved the regular distribution of dye molecules throughout the entire film thickness. The low internal diffusion time seems to be responsible for preventing non-uniformity problems along the film thickness. In fact, a high dye uptake rate represents a certain advantage since the total dyeing time can be shortened ten times as compared to conventional dyeing, whereas, faster dye uptake kinetics is apt to cause non-uniform dye distribution in the final product when dyeing is done traditionally.

7.3.5. Temperature Effects

The dyeability was found to strongly depend on the dye bath temperature in the plasma dyeing process, as shown in Figure 5b; this is also a common phenomenon in traditional dyeing processes.[13] Results show that the amount of dye absorbed on the coating decreases with decreasing dye-bath temperature. The low reduction in dyeability at lower temperature (<80°C) is probably due to dye aggregation and a low degree of adsorption. The differences, however, became very small especially at higher temperatures within the range of 80-120°C. By increasing the temperature, dye uptake can be enhanced because of the increased solubility and mobility of dye molecules in water. There may be another reason for this finding that, which is commonly seen in traditional synthetic dyeing processes: the enhanced color yield, which gradually increases with increasing temperature, can be attributed to a corresponding increase in the amount of accessible volume available for dye diffusion according to the "free volume model".[14,15] The highest color yields were found to be at ≈100°C implying that the dye uptake reached the "saturation level" resulting in maximum acid-base intermolecular interaction between dyes and amine functionalities. At this level, the dye molecules occupied most of the incorporated amine functionalities in the film volume yielding maximum dyeability. Thus, substrates such as PES, PP, aramid, glass textiles etc. can be dyed at low temperatures similar to wool dyeing, since good coloration was obtained at levels as low as 80°C.

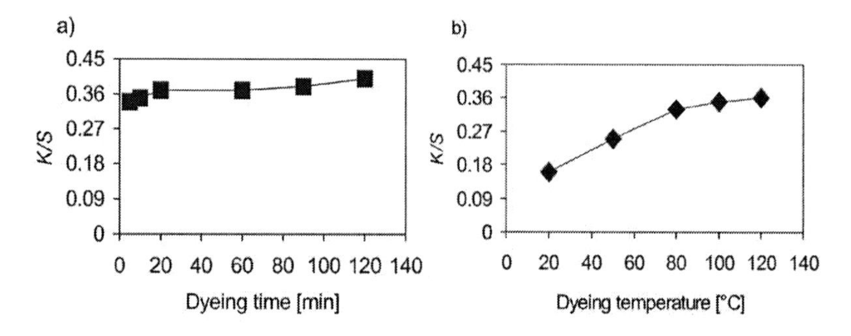

Figure 5. Effect of dyeability on dyeing parameters: a) dyeing time (80°C), and b) dyeing temperature (60 min) of plasma-coated (600 W, NH_3/C_2H_4 = 0.84, 20 min) and light-shade (0.5% o.w.f.) dyed PES.

7.3.6. Dyebath pH

The pH value of the dye bath was found to be very important in order to achieve level dyeing and to increase dye affinity to the functionalized films. The optimum level dyeing and color strength were obtained at a pH in the range of 4.5-5. However, the color intensity was increased at pH 2.5-3 due to improved exhaustion, but uneven shade was observed. On the other hand, dyeability was reduced at pH 6.0-7.0 due to low substantivity or low affinity of dyes to the film resulting in weak dye-film interaction.

7.3.7. Wash Fastness

Wash fastness at 60°C was evaluated according to the EN ISO 105-C06 standard, which is summarized in Table 1. Wash fastness was examined for the samples, which showed the highest color strength values obtained. It can be seen that satisfactory wash fastness was obtained indicating a permanent ionic bond between dyes and amines in the films. The results confirmed decreased dye reduction and migration behavior.[16] The homogeneous nanoporous coating with a crosslinked and branched structure enhanced uniform dye molecule access into the a-C:H:N films. As a consequence, uniform and level dyeing was obtained.

Table 1. Wash fastness at 60°C of plasma-treated (600 W, 20 min) and dyed PES textiles.

NH_3/C_2H_4 ratio	Wash fastness		
	Color change	Staining (PES)	Staining (wool)
0.71	3	5	3-4
0.83	3	5	3-4
1.0	3	5	3
1.25	3	5	3

7.4. Conclusion

Nanoscaled ultrathin films (<80 nm) were deposited on PES fabrics by ammonia/ethylene plasma treatments which resulted in the incorporation of polar amines containing functional groups. Nitrogen incorporated a-C:H films facilitated stable and durable nanoporous crosslinked surfaces. The nature of the coatings varied strongly depending on the gas ratio and energy input. While higher ammonia to monomer ratios showed a more dense film structure indicated by a reduction in dyeability, lower gas ratios (NH_3/C_2H_4 = 0.84-1.0) revealed excellent dye uptake representing high amine functionalities inside the porous coatings. Thus, the nanoporous structure provided a high surface area that facilitated the incorporation of amine functional groups. Coloration of deposited a-C:H:N films demonstrated substrate independent superficial dyeing e.g. PES, PP, aramid, glass etc., since nitrogen functions formed in the a-C:H:N films can easily be accessed by dye molecules. The color strength value can be improved by varying plasma dyeing parameters. Homogeneous distribution of functionalities and voids in the coatings yielded level dyeing even in a very short time. The short dyeing time ensured faster and maximum penetration/diffusion of dyes into the coatings. Therefore, dyeing time does not significantly influence color intensity in plasma dyeing process, thus dyeing time could be reduced greatly. A very good color yield was achieved aready at $\approx$80°C indicating the (almost) maximum degree of exhaustion. The introduced polar dye sites became saturated at above 120°C. Good wash fastness confirmed the permanency of dye-film bonds. Both ammonia/acetylene and ammonia/ethylene plasmas facilitate the deposition of nitrogen containing functional coatings, which are accessible to dye molecules throughout the film volume, as described in the chapters (6-7). In addition, an extended study was performed as shown in the following chapter in order to improve the coating quality with an increased functional density.

7.5. References

[1] M. M. Hossain, A. S. Herrmann, D. Hegemann, *Plasma Process. Polym.* **2007**, *4*, S1068.
[2] D. Hegemann, *Indian J. Fibre Text.* **2006**, *31*, 99.
[3] D. F. Franceschini, *Braz. J. Phys.* **2000**, *3*, 517.
[4] N. K. Cuong, M. Tahara, N. Yamauchi, *Surf. Coat. Technol.* **2005**, *193*, 283.
[5] M. Creatore, P. Favia, G. Tenuto, A. Valentini, d`Agostino, *Plasma. Polym.* **2000**, *5 (3/4)*, 201.
[6] E. Riedo, F. Comin, J. Chervier, *J. Appl. Phys.* **2000**, *88*, 4365.
[7] D. Hegemann, M. M. Hossain, D. J. Balazs, *Prog. Org. Coat.* **2007**, *58*, 237.
[8] M. Creatore, P. Favia, G. Tenuto, A. Valentini, R. d'Agostino, *Plasma Polym.* **2000**, *5 (3/4)*, 201.
[9] K. M. Siow, L. Britcher, S. Kumar, H. J. Griesser, *Plasma Process. Polym.* **2006**, *3*, 392.
[10] M. M. Hossain, A. S. Herrmann, D. Hegemann, *Plasma Process. Polym.* **2007**, *4*, 135.
[11] D. Hegemann, M. M. Hossain, *Plasma Process. Polym.* **2005**, *2*, 554.
[12] M. Banchero, A. Ferri, *J. Supercrit. Fluid.* **2005**, *35*, 157.
[13] A. R. T. Bagha, H. Bahrami, B. Movassagh, M. Arami, F. M. Menger, *Dyes Pigments* **2007**, *72*, 331.
[14] "*Chemical Principles of Synthetic Fibres Dyeing*", S. M. Burkinshaw, Eds., Blackie, Glasgow, **1995**.
[15] T. Kim, Y. Son, Y. Lim, *Dyes Pigments* **2005**, *67*, 229.
[16] Y. A. Son, J. P. Hong, T. K. Kim, *Dyes Pigments* **2004**, *61*, 263.

8. Comparison Study Between NH_3/C_2H_2 and NH_3/C_2H_4 Discharges

8.1. Introduction

Over the past decade, functionalized coatings have been proven to have great importance in materials science for high value materials. The incorporation of functional groups on the substrate surface alters radically many important properties such as hydrophilicity, dyeability, adhesion, friction etc.[1-7] The properties of synthetic polymers may be suitably modified and tailored during synthesis; however it may influence the physical properties such as comfort and touch, porosity, and air/water permeability for fibrous materials, in particular textiles. Moreover, the use of additional chemicals such as solvents, surfactants, reducing agents etc. may affect the intrinsic bulk properties of the materials. Ecological concerns have resulted in a renewed interest in dry-systems. Environmental safety and ecological aspects are becoming increasingly important for the introduction of new materials, and plasma technology meets these criteria better than wet chemistry processes.

Plasma deposition has been used extensively for the fabrication of thin films due to their high potential in technological applications.[8,9-12] The modification has two major advantages: one of them is the removal of manufacturing residuals and contaminations present on the fabrics by combustion reactions and micro-ablation using Ar/O_2 plasma.[13,14] The other is to deposit nanoscaled coating with a desired surface chemistry by incorporation of functional groups on the outermost fabric surface using ammonia/acetylene and ammonia/ethylene plasmas.[15,16] Even though nitrogen-containing plasma systems (non-polymer forming gases such as N_2, NH_3, O_2/NH_3, Ar/NH_3 etc.) have been extensively used in order to introduce amine, imine, amide, cyanide, nitrile etc., functionalities to the polymer surfaces, the modification was not durable for longer time periods due to restructuring of the functional groups and/or the chain migration in the surfaces, as described in numerous references.[17-21] These non-permanent modifications could be used for subsequent processes such as lamination, coloration etc.; however it has strong limitation in industrial applications. Ammonia with hydrocarbon gaseous mixtures allows one to obtain nitrogen-incorporated amorphous hydrocarbon (a-C:H:N) films. The deposited films described here strongly reduce hydrophobic recovery and preserve nitrogen-functionalities in the coating for a longer time. Therefore, this work investigates the generation of permanent functional coatings. As the coating contains polar functions, it alters the surface energy significantly from hydrophobic to hydrophilic. The addition of nitrogen contributes to nanoporous and crosslinked coatings by rivaling deposition/etching processes. As a result, functionalities in the coating facilitate corresponding dye-molecule access and form a dye-fiber bond. Thus, nanoporous a-C:H:N films help to obtain a substrate independent surface dyeing representing the characterization of chemical surface properties.

Another important goal of this study was to achieve a uniform deposition and to characterize the homogeneity of the functional groups distributed in the plasma polymers. Much attention therefore was given to the development of thin coatings that resulted in uniform distribution of functionalities on the substrate being deposited. XPS may provide the chemical composition of the plasma nanofilm including reactive functionalities within the film and elemental surface composition etc. However, it can influence the chemical composition such as the loss of chemical information, for example, changes and/or decomposition of the functional groups (preferential sputtering) during in-depth profiling by an ion beam for the sputter erosion of the surface. Moreover, the characterization of deposited film (<100 nm) on the fabrics is very difficult because of their complex geometries compared to flat and solid polymer surfaces. Therefore, the coloration of the deposited films on textiles provided useful information about the coating uniformity, purity, and the specific surface chemistry.

In this chapter, NH_3/C_2H_2 and NH_3/C_2H_4 plasmas were investigated to a great extent in order to enhance accessible amine functionalities and to reduce unsaturated bonds in the plasma deposition. The results of functional coatings and the coloration were compared for both plasmas. The a-C:H:N films were deposited on fabrics by cold plasma using a pilot plant plasma reactor. The deposited-hydrophilic a-C:H:N films were characterized by contact angles (CAs). The mechanical stability of the plasma coating was examined using an Abrasion & Pilling Tester. Dyeing of the plasma coating was examined with a Datacolor Spectraflash, while a study was done to observe the influences of energy input *W/F* (J/cm^3), film thickness, and gas ratio.

8.2. Experimental

8.2.1. Functionalized Plasma Coating

Empa pilot-scale web coater and its vacuum system are elaborately described in Chapter 6.[16] The web-coater is able to treat textiles, foils, membranes, papers etc. semi-continuously with a maximum width of 65 cm. When a selected gaseous mixture is introduced into the vacuum chamber, a capacitively coupled RF plasma was used to activate and excite the gas molecules to produce metastable plasma species such as radicals, ions, electrons, energetic particles, and VUV. Due to the symmetric electrode set-up – the RF driven drum, where the fabrics are processed, has a similar size compared to the chamber wall – the substrate temperature remained close to the room temperature due to the low bias voltage.

A two-step plasma process was carried out in order to achieve a permanent functionalized polymer coating on the textiles. The process consists of a non-polymer forming plasma for cleaning, activation and roughening of the fiber surfaces, and subsequent deposition of a functionalized nanoscaled film. Firstly, Ar/O_2 plasma treatment was carried out for one minute at 400 W and 10 Pa gas pressure as pretreatment of the textile samples. Secondly, a functional plasma polymer was deposited on the cleaned and nano-textured fabric surfaces using ammonia containing acetylene and ethylene plasma in a systematic way by examining different energy inputs by the reaction parameters *W/F* and gas ratios (NH_3/C_2H_2 and NH_3/C_2H_4).

8.2.2. Contact Angle Measurement

CA measurements were performed according to the conditions described in Chapter 6.

8.2.3. Coating Permanency and Hydrophilicity

Deposition of functionalized nanoscaled coatings on the fiber surface has been found to be very efficient in order to obtain a permanent hydrophilic modification as compared to non-polymer forming plasmas. Due to the covalent bonding and their polar chemical nature, the functionalized films adhered well to the substrate which was examined by adhesion and abrasion tests. The coated fabric surface was found to be smooth, which reduced pilling and improved abrasion resistance of the fabrics. Moreover it may be possible to reduce cracks, flaws and fiber irregularities on the fabrics by controlling the plasma film thickness.[22] As long as the coating thickness is in the nanometer range (<100 nm), the fabric touch and comfort was not affected. Moreover, it may slightly increase the tensile strength of the fabric.[16]

8.2.4. Dyeing Procedure

The dyeing procedure and the substrate independent dyeing mechanism[23,24] were described in Chapter 6.

8.2.5. CIELAB Color Coordinates

CIELAB color spaces describe a three dimensional color system, where a^*, b^*, L^*, C^* represent red-green axis, blue-yellow axis, darkness-lightness axis, color brilliance, respectively, and h denotes hue of a color (in deg.). Total CIELAB color difference was measured using the following formula where the plasma-treated PES was considered as the reference fabric:

$$DE^* = \sqrt{\{(Da^*)^2 + (Db^*)^2 (DL^*)^2\}} \qquad (8.1)$$

Here, DE^* = total color difference, Da^* = color difference in redness/greenness, Db^* = color difference in yellowness/blueness and DL^* = color difference in lightness/darkness. A depletion of b^* value tends to be blue color axis and a positive b^* value drifts to be yellow color axis.

8.3. Result and Discussion

8.3.1. Effect of Monomer Gas on Plasma Polymerization

In order to observe the deposition rate of the polymer-forming hydrocarbon gases (a-C:H coatings), experiments were carried out with pure monomers such as acetylene and ethylene. In all cases, deposition in mass was measured by weighing the deposited polymer per square area per unit time (g/min cm^2), and then normalized by the corresponding monomer flow (sccm) to evaluate the growth of plasma polymers.[25] Depending on the process parameters (specifically monomer flow and gas pressure) plasma polymer formation varied strongly regarding uniform coating or powder formation. Especially with increasing pressure (>5 Pa), powder (dirt particles) was generated in pure hydrocarbon plasmas and deposited on the substrate surface in particular for C_2H_2 plasma as compared to C_2H_4.[26,27] On the other hand, at reduced pressure (2.5 Pa), uniform a-C:H coatings were obtained by varying the energy input ($W/F \approx$240-480 J/cm^3). Therefore, monomer flow and gas pressure should be carefully considered, in order to obtain a uniform and stable coating and/or to enhance adhesion between film and substrate surface.

Figure 1 shows that plasma film growth is strongly influenced by the specific energy input *W/F*.[28] In the Arrhenius-type plot the deposition rate of plasma polymers depends linearly on the inverse energy input $(W/F)^{-1}$. The fragmentation of acetylene or ethylene within the glow discharge initiates plasma polymerization. It includes mainly two processes: the detachment of hydrogen from the monomer and the bond scission of the unsaturated triple or double bond yielding mono- and divalent radicals, respectively. As a result, the plasma polymer formation on the substrate increases at higher energy inputs due to the enhancement of radical formation in the active plasma zone. As shown in Figure 1, C_2H_2 plasmas revealed a slightly higher mass deposition rate compared to C_2H_4 plasmas for similar plasma conditions. $HC{\equiv}CH$ needs less energy (2.0 eV) to open one of the triple bonds (whereas 5.5 eV is required to abstract hydrogen), while $H_2C{=}CH_2$ requires more energy (2.8 eV) to open its double bond, i.e. the formation of divalent radicals is easier. As a result, C_2H_2 can probably produce a higher radical yield compared with C_2H_4 plasmas resulting in a slightly increased deposition rate. The higher number of hydrogen atoms in the $H_2C{=}CH_2$ system might further cause a reduction in deposition rate through chemical etching as compared to a $HC{\equiv}CH$ system.[29]

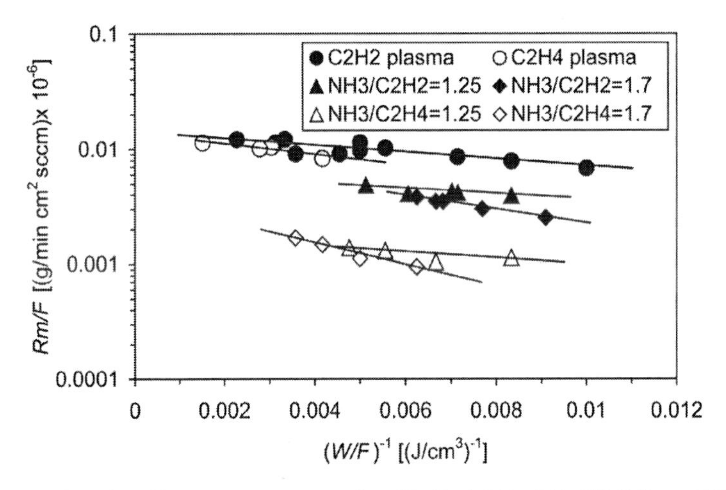

Figure 1. Mass deposition rates per monomer flow (*Rm/F*) of C_2H_2, C_2H_4, NH_3/C_2H_2 and NH_3/C_2H_4 discharges as a function of the inverse of energy input *W/F*.

8.3.2. Effect of Ammonia in the a-C:H:N Deposition

As mentioned above, the deposition rate was measured based on hydrocarbon monomers, as it participates directly in film growth, whereas the admixture of non-polymerizable gases like ammonia yield etching/co-deposition processes. Figure 1 shows the growth rate of plasma polymers, when ammonia was introduced in the reaction chamber. It can be clearly seen that the growth rate has decreased by a factor of 2-3 for the NH_3/C_2H_2 plasma and a factor of 5-6 for the NH_3/C_2H_4 plasma compared to pure a-C:H film growth. Note that the specific energy *W/F* in Figure 1 is related to the monomer gas, while some energy is also consumed by the non-polymerizable gas.[15] However, it clearly demonstrates that the addition of ammonia in the plasma atmosphere causes etching of a-C:H film. Pure ammonia plasma is known to induce very strong etching effects due to its highly reactive amines or nitrogen-contained species.[30] Y. L Zou[31] reported that pure ammonia leads to a significant reduction in tensile strength for graphite fibers over prolonged plasma exposure times. On the other hand, etching processes of ammonia mixed with a polymerizable gas were found to be dependent on the ammonia/monomer ratio at a constant pressure of 10 Pa, which should not affect the fiber strength.[31] Hence, simultaneous etching and polymerization within an ammonia/hydrocarbon discharge can be used to deposit functional coatings such as a-C:H:N films. Etching or texturing of the substrate contributes to higher mechanical interlocking of the plasma polymers yielding an enhanced interfacial adhesion. Moreover, a crosslinked and branched hydrocarbon structure containing nitrogen functionalities such as amino groups within the a-C:H:N films can be generated, which is a prerequisite in order to get stable and uniform nanoporous coatings. Plasma pretreatment can further improve the mechanical stability of the coatings. It was proved that activation/cleaning was remarkably enhanced using noble gases such as Ar or He mixed with O_2, as shown in Chapter 4. [13,32] Thus, an Ar/O_2 plasma was used to activate polyester (PES) fabrics.

At the gas ratios used (NH_3/C_2H_2 or NH_3/C_2H_4) deposition was enhanced as the energy input increased due to more molecular fragmentation and radical formation. The deposition rate varied depending on the ammonia/monomer flow ratio, while other plasma parameters remained constant (Figure 1). Within the range of parameters examined, higher ammonia flow contributed to more nitrogen incorporation within the coating, while higher monomer flow yielded higher film growth rates for both of the hydrocarbon monomers used. The number of accessible amine functionalities can thus be controlled by film thickness and density of the groups depending on the application area. In surface dyeing, for example, the color strength strongly depends on both parameters, as shown below, while in the case of composite textiles the improvement of fiber/matrix adhesion by a-C:H:N films is less influenced by film thickness. Higher film growth might be obtained at a higher pressure; however, the coating stability might be a critical factor. At a higher plasma pressure (480 Pa) ammonia/ethylene-derived films were found to change their nature from polymer films to powder-like products.[27]

It is remarkable that the deposition rate in NH_3/C_2H_2 plasmas was distinctly higher compared to NH_3/C_2H_4 plasmas, while both hydrocarbons in pure monomer discharges showed a similar film growth (see Figure 1). Hence, a change in the dominating plasma polymerization reactions might be the reason. Bond scission between carbon atoms in the hydrocarbon molecules yielding divalent radicals leads to higher deposition rates compared to hydrogen abstraction. This can be assumed to be the main polymerization mechanism for both acetylene and ethylene, since a strongly reduced deposition rate for pure methane plasmas was observed.[16] Admixture of ammonia influences the plasma polymerization both by gas phase reactions and etching at the surface, which might be stronger for ethylene molecules. Moreover, a higher hydrogen yield in the C_2H_4/NH_3 plasma might restrict the plasma polymer formation.

8.3.3. Contact Angle Measurement and Surface Wettability

Chemical functionalities and roughness of the plasma formed polymer surface can dominantly be explained by CAs. Figures 2-3 show static CAs of plasma treated and untreated glass substrates as a function of aging for different gas ratios of NH_3/C_2H_2 and NH_3/C_2H_4, respectively, at energy inputs between 140 and 420 J/cm^3 (increasing for higher gas ratios) and a constant gas pressure of 10 Pa. In all cases, the CAs were reduced remarkably compared to untreated PES (CA around 80°) depending on the chemical composition at the surface and morphological changes.

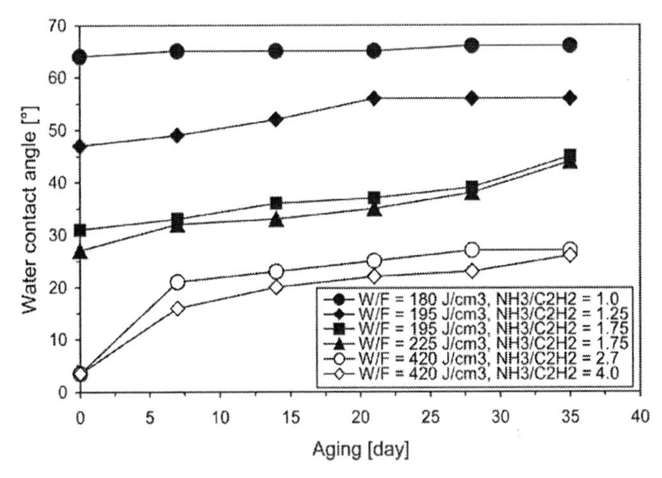

Figure 2. Static water contact angles on NH_3/C_2H_2 plasma-treated glass substrate with aging for different energy inputs *W/F* and gas ratios.

As can be observed in Figures 2-3, CAs changed strongly depending on the gas flow ratio, whereas energy input was found to have merely a minor influence. High ammonia flow with respect to the monomer gas contributed to the formation of a high number of nitrogen functionalities such as amines, amides, cyanides, imines etc.[33,34] on the plasma polymers surface indicating a strong reduction in CA. Hydrophilicity strongly depends upon surface energy due to incorporation of polar functional groups.[35] It is interesting to find that NH_3/C_2H_2 flow ratio$\geq$2.7 showed a very strong hydrophilicity (CAs<25°) for ammonia in the acetylene plasma, whereas already lower gas ratios, i.e. NH_3/C_2H_4 flow ratio$\geq$1.7, yield permanent hydrophilicity (CAs<16°) for ammonia in the ethylene plasma at similar energy inputs for both plasmas, probably due to a texturing of the film surface which supports capillary effects.[36,37] Morphological modification on the surface such as surface roughness, surface etching, texturing etc. contributed to a superhydrophilic surface when plasma additionally produced hydrophilic functional polar groups. A higher ammonia content in the acetylene or ethylene atmosphere caused chemical and physical etching effects, therefore, significant changes in CAs were found at higher gas ratios for both plasmas. Inagaki found similar results with ammonia plasma (50 W, 13.3 Pa) on the surface roughness of PET films.[38] Hence, the hydrophilicity not only depends on surface energy by incorporation of polar functional groups, but also on the morphological changes of the plasma modified surface.

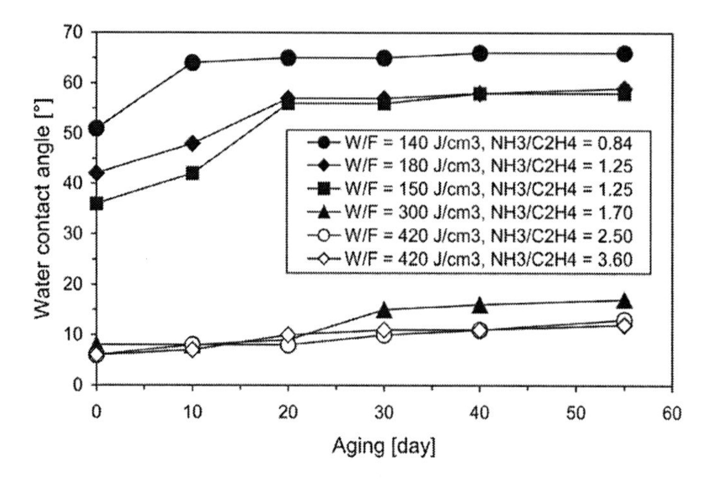

Figure 3. Static water contact angles on NH_3/C_2H_4 plasma-treated glass substrate with aging for different energy inputs *W/F* and gas ratios.

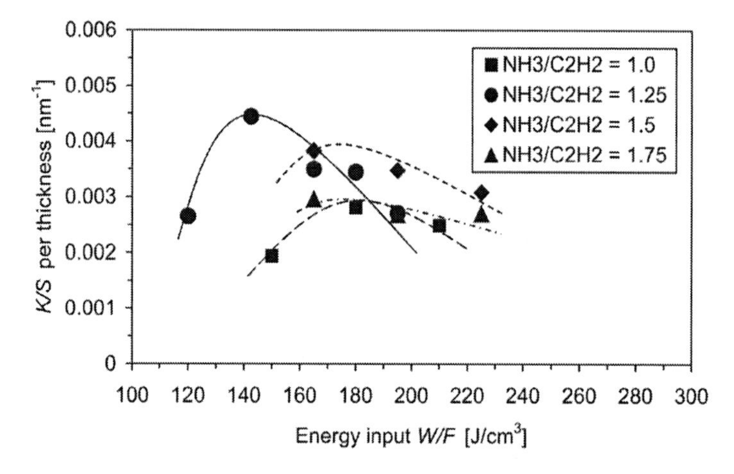

Figure 4. Color strength values (*K/S*) of NH_3/C_2H_2 plasma-treated and dyed PES textiles depending on the film thickness and energy inputs *W/F*.

In addition, it can be assumed that etching does not change remarkably with energy input *W/F*, since the deposition rates were found to increase with *W/F* for a fixed gas ratio. Moreover, increasing ammonia content might yield more dangling bonds and saturation with oxygen functionalities at the surface. Aging, i.e. hydrophobic recovery, was found to take place mainly within the first two weeks and proceeded slowly over the next several weeks for both plasma systems, and was caused mainly by surface adsorptions. The change in CAs with aging is caused by the restructuring of polar groups on the coated substrate. After two weeks of aging CAs became constant, indicating that the hydrophilic treatment is permanent for longer time periods. Hence, both acetylene and ethylene mixed with ammonia are suitable to achieve durable hydrophilic textile surfaces.

8.3.4. Accessing Amine Functionalities by Acid Dyestuffs

Since only the plasma-polymerized coating is dyed, the coloration strongly depends on the film thickness and density of accessible nitrogen functional groups inside the coating. Nitrogen functionalities, mainly amino group containing plasma polymers,[32,39,40] facilitate the superficial dyeing of the PES substrate using acid dyestuffs. Coloration of PES with this dye class is generally impossible, since untreated PES does not contain amine groups in their structure needed for the dye-fiber reaction mechanism (see Chapter 6). Figures 4-5 demonstrate the color intensity given by *K/S* values normalized by film thickness of the plasma-treated and dyed PES as a function of energy input *W/F* and gas ratio. It illustrates that color intensity is correlated both to gas ratio and energy input. Experiments were done in a wide range of gas ratios. However, only the ratios yielding high color intensities are presented. Higher ammonia to hydrocarbon ratios led to more etching during plasma polymerization, as indicated by a reduced deposition rate. These conditions might reduce the density of accessible amine functional groups within the coatings, since color strength per thickness was found to be very small (<0.002/nm), although a high flow of ammonia in the monomer gas ($NH_3/C_2H_2 \geq 2.7$ and $NH_3/C_2H_4 \geq 1.7$) was found to produce polar groups at the surface as indicated by low CAs (<25°).

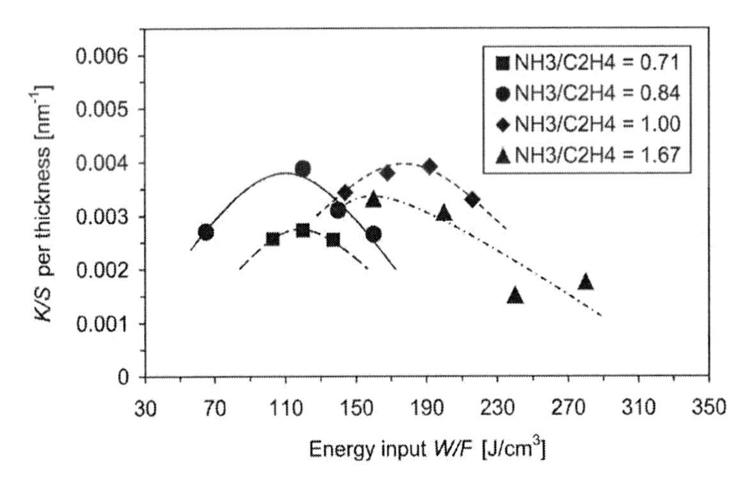

Figure 5. Color strength values (*K/S*) of NH_3/C_2H_4 plasma-treated and dyed PES textiles depending on the film thickness and energy inputs *W/F*.

Therefore, it can be assumed that coloration does not only depend upon the wetting properties given by surface functionalities and texturing, but strongly depends on the amount of accessible functional groups within the coating, which can chemically bind acid dye molecules. Thus, a nanoporous, mechanically stable a-C:H:N coating with a desired number of functional groups inside the film guarantees a high coloration rate,

where the dye molecules are small enough (about a few nanometer) to penetrate easily through interconnected voids of the nanoporous structure of the polymer film and form dye-film bonds.
A high deposition rate at a low ammonia flow rate in the monomer gas, on the other hand, yields low coloration due to the insufficient number of nitrogen functional groups in the plasma polymer. In this range, CAs were found to be high (CAs ≥ 60°) due to the predominant hydrocarbon formation during deposition. Thus, CAs are not directly linked to plasma-assisted dyeing and amine functionalities. Therefore, the ammonia flow in the monomer gas should be in the range where both incorporation of nitrogen functionalities and formation of a nanoporous structure can take place. An optimum combination assures high color strength. It is found that at a ratio of $NH_3/C_2H_2 = 1.25$ the best color intensity rate (0.0045 *K/S* per nm) was observed in the acetylene atmosphere, while a NH_3/C_2H_4 ratio of 0.84 showed the best color intensity rate (0.0043 *K/S* per nm) for the ethylene atmosphere. The results imply that for both plasmas similar color strength values can be obtained at slightly different gas ratios. Some differences between acetylene and ethylene mixed with ammonia were already discussed in the film growth section showing reduced deposition rates for NH_3/C_2H_4 indicating a different polymerization mechanism.[41]
Energy input *W/F* is another important factor to achieve high *K/S* values. At rather low energy input *W/F* a reduced dyeability was observed caused by low molecular fragmentation. At moderate *W/F* in the range of 140-180 J/cm^3 for NH_3/C_2H_2 plasmas and 120-170 J/cm^3 for NH_3/C_2H_4 plasmas excellent *K/S* values were obtained as can be seen in the Figures 4-5. The results proved that these energy ranges assured an optimum combination of functionalization and deposition indicated by a maximum incorporation of reactive nitrogen functionalities inside the coating. Furthermore, dyeability was found to decrease again at around $W/F > 180$ J/cm^3 for both plasmas indicative of strong fragmentation yielding less accessible nitrogen functionalities in the coating that can attach dye molecules.

8.3.5. Improved Coating Quality by NH_3/C_2H_4 Plasma

Coating quality and purity can be examined in respect of the yellow color component in light-shade blue-color dyeing. Figure 6 evidently demonstrates that surface yellowness in the plasma-coated and dyed PES can be decreased remarkably using ammonia/ethylene plasma as compared to ammonia/acetylene plasma, while the total color difference remained almost identical. As the yellowness was decreased, the blueness of the dyed sample was enhanced accordingly. It can be concluded that ammonia/acetylene plasma polymers contain more unsaturated bonds compared to ammonia/ethylene plasma, which predominantly caused yellow color by light absorption. One reason might be the higher bond strength of the triple bond in the $HC{\equiv}CH$ system compared to the double bond in the $H_2C{=}CH_2$, which makes the retention of such unsaturated bonds within the plasma polymer more probable. Besides, the $H_2C{=}CH_2$ system might form a more saturated hydrocarbon network mainly by opening the double bond, revealing different polymerization mechanisms for NH_3/C_2H_2 and NH_3/C_2H_4 plasmas, which was also indicated by the difference in deposition rate.

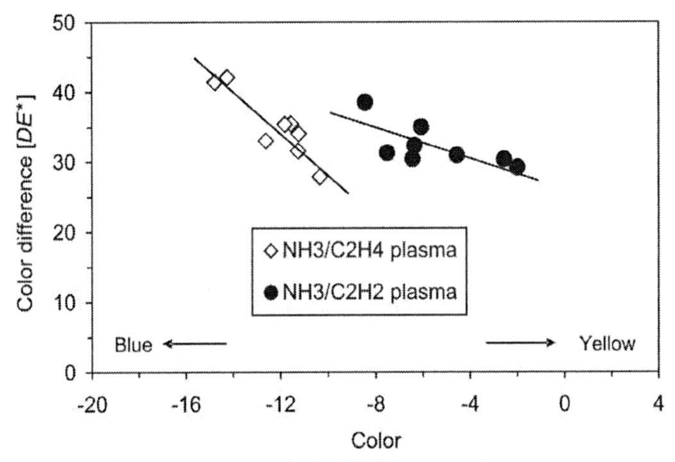

Figure 6. Reduced yellow color component in the NH_3/C_2H_4 atmosphere.

Tables 1-2 show the CIELAB color coordinates as a function of gas flow ratio and specific energy *W/F*. The results are compared using the optimum plasma parameters with respect to dyeability. It is apparent that the hue of ammonia/ethylene treated and dyed PES was improved greatly compared to ammonia/acetylene as the yellowness index was reduced to a great extent. Though lightness/darkness values stay roughly unchanged for both plasmas, chroma was enhanced for ammonia/ethylene plasma due to improved color brilliance. Thus, the coating quality and purity can be enhanced using an ammonia/ethylene plasma atmosphere.

Table 1. CIELAB color coordinates of NH_3/C_2H_2 plasma-treated (10 Pa) and dyed PES textiles.

NH_3/C_2H_2 ratio	*W/F* (J/cm^3)	*L**	*h°*	*a**	*b**	*C**
1.0	180	68.04	193.1	-11.0	-2.56	11.3
1.0	210	65.83	205.7	-12.6	-6.07	14.0
1.25	195	68.36	190.0	-11.3	-1.99	11.5
1.25	225	62.30	212.9	-13.1	-8.44	15.6
1.5	195	67.89	208.6	-11.7	-6.36	13.3
1.5	225	68.11	202.0	-11.2	-4.55	12.1
1.75	195	69.18	213.0	-11.6	-7.52	13.8
1.75	225	69.25	210.3	-11.0	-6.44	12.8

Table 2. CIELAB color coordinates of NH_3/C_2H_4 plasma-treated and dyed PES textiles.

NH_3/C_2H_4 ratio	*W/F* (J/cm³)	*L**	*h*°	*a**	*b**	*C**
0.71	100	71.38	225.9	-12.2	-12.61	17.6
0.71	120	74.22	222.5	-11.3	-10.34	15.3
0.71	140	72.01	223.2	-12.0	-11.25	16.4
0.83	120	68.64	228.2	-10.1	-11.23	15.1
0.83	140	68.62	221.8	-12.9	-11.55	17.3
0.83	160	69.09	223.9	-12.3	-11.81	17.0
1.0	170	65.40	228.2	-13.2	-14.76	19.8
1.0	190	64.02	226.2	-13.6	-14.24	19.7

Table 3. Wash (at 40°C) and rub fastness of NH_3/C_2H_2 plasma-treated and dyed PES textiles.

Plasma parameters		Wash fastness			Rub fastness	
NH_3/C_2H_2 ratio	*W/F* (J/cm³)	Color change	Staining (PES)	Staining (wool)	Dry	Wet
1.0	180	3	5	4	4-5	4-5
1.0	210	3	5	4	4	4
1.25	195	3	5	4	4	4
1.25	225	3	5	4	4	4-5
1.5	195	3	5	4	4	4-5
1.5	225	3	5	4	4	4
1.75	195	3	5	4	3-4	3-4
1.75	225	3	5	4	3-4	3-4

Table 4. Wash (at 60°C) and rub fastness of NH_3/C_2H_4 plasma-treated and dyed PES textiles.

Plasma parameters		Wash fastness			Rub fastness	
NH_3/C_2H_4 ratio	*W/F* (J/cm³)	Color change	Staining (PES)	Staining (wool)	Dry	Wet
0.71	100	3	5	3-4	4	4-5
0.71	120	3	5	3-4	3	4-5
0.71	140	3	5	3-4	4	4-5
0.83	120	3	5	3-4	3	4-5
0.83	140	3	5	3-4	4	4-5
0.83	160	3	5	3	4	4-5
1.0	170	3	5	3	3	4-5
1.0	190	3	5	3	2-3	4

8.3.6. Permanency and Uniformity of Plasma Film

Tables 3-4 show summarized results of plasma-dyed PES which were assessed using an ISO test method (ISO 105-X12 for color fastness to rubbing and EN ISO 105-C06 for color fastness to washing). Under D65 illumination color changes, staining and rub were evaluated using grey scales: ISO-105-A02 grey scale for assessing change in color; ISO-105-A03 grey scale for staining and rub. Both plasmas show almost comparable fastness properties after washing at 40°C (Table 3) and 60°C (Table 4) for ammonia/acetylene and ammonia/ethylene plasmas dyed PES respectively. This means that ammonia/ethylene plasma has better wash fastness than ammonia/acetylene plasma, since a higher temperature could be used. The acid dyeing of a-C:H:N plasma polymers exhibit acceptable fastness to laundering and rubbing. From these results, it can be concluded that a nanoporous and functional plasma polymer enabled a permanent dye-film interaction and demonstrated stability and longevity of amine groups in the coating.
The coating was found to adhere well to the textile surface. No damages to the coating were detected after 60,000 rubbing cycles. The surface activation was determined to be efficient to obtain the mechanically stable coating due to the enhanced chemical bonding and adhesion by surface cleaning and addition of polar functionalities. As the coloration results showed, excellent even dyeing was obtained. High quality uniform dyeing guaranteed equal and even distribution of amine functionalities throughout the entire film thickness.

8.4. Conclusion

Ammonia/hydrocarbon mixtures were investigated within a RF plasma using a web coater in order to deposit a-C:H:N coatings on PES textiles. Pure hydrocarbon discharges at low pressures can be used to deposit crosslinked a-C:H coatings. In particular, high deposition rates can be obtained with unsaturated hydrocarbon monomers such as acetylene and ethylene by producing divalent radicals. Admixture of ammonia to the hydrocarbon discharge influenced the plasma polymerization mechanism depending mainly on gas ratio and energy input. Etching and co-polymerization processes yielded a reduction in the deposition rate and the incorporation of nitrogen functional groups, which was found to be more pronounced for NH_3/C_2H_4 plasmas. In addition, nitrogen incorporation in a-C:H films facilitated the formation of crosslinked and branched plasma polymers. Thus, nanoporous and uniform coatings can be obtained.
While increasing ammonia content caused a drop in the deposition rate, a strong reduction in CAs proved the formation of polar groups on the nanoscaled plasma coatings, which were found to be durable with storage time. The reactive ammonia caused texturing and roughening of the surfaces by etching of plasma polymers, which contributed to enhanced surface wettability and hydrophilicity. Coatings deposited at gas ratios of $NH_3/C_2H_2 \geq 2.7$ and $NH_3/C_2H_4 \geq 1.7$ enabled a permanent hydrophilization, i.e. complete wetting, of textile fabrics such as PES.
The coloration of a-C:H:N films on hydrophobic PES textiles proved that the nanoporous coating facilitated the access of hydrophilic acid dyes to nitrogen groups in the coatings, which yielded substrate independent surface dyeing. Higher gas ratios were found to yield less coloration due to fewer amine functional groups, while they show higher hydrophilicity owing to the addition of polar functions on the surfaces during post plasma reactions. At gas ratios of around 1.25 for NH_3/C_2H_2

and 0.84 for NH_3/C_2H_4 crosslinked and branched a-C:H:N coatings were produced, which composed of a high number of accessible nitrogen functional groups, i.e. amino groups, resulting in high color intensity. Although their wetting properties were less pronounced than for higher ammonia/hydrocarbon content, dye molecules could easily enter the nanoporous film structure and chemically bond with the nitrogen functionalities. Thus, high *K/S* values per film thickness were obtained. Moreover, the plasma-deposited and dyed PES fabrics showed a good rubbing and washing fastness demonstrating the coating-functional permanency. The excellent abrasion resistance confirmed that the coating was permanently adhered to the substrate.

While plasma polymers deposited from NH_3/C_2H_2 mixtures showed a noticeable yellow color component with respect to the CIELAB color system, the yellowness can be strongly reduced using ethylene instead of acetylene. Hence, plasma polymers from acetylene mixtures retain unsaturated carbon groups within their structure, while the double bond of ethylene is more or less opened during plasma polymerization. Slightly different polymerization mechanisms can thus be assumed for NH_3/C_2H_2 and NH_3/C_2H_4 plasmas yielding differences in deposition rate, wetting properties and dyeability as well as yellowness index depending on the gas ratio used. Comparable results can be obtained for both plasmas when lower amounts of ammonia are used for ethylene compared to acetylene. Therefore, the coating quality and purity can be improved significantly using an ammonia/ethylene gaseous mixture.

These findings introduce new advanced functionalities in fabrics which can be potentially used in various fields such as the incorporation of bio-molecules in biotechnological applications, protein adsorption in medical applications, improvement of interfacial adhesion and mechanical interlocking in fiber-reinforced composites, and UV absorption in protective clothing etc. while preserving strong hydrophilicity as described in details in the next chapter. New added functions in fabric systems are desirable for the next generation of wearable chemical and biological protection.

8.5. References

[1] M. M. Hossain, A. S. Herrmann, D. Hegemann, *Plasma Process. Polym.* **2007**, *4*, 135.
[2] T. Wakida, S. Tokino, *Indian J. Fibre Text. Res.* **1996**, *21*, 69.
[3] H. Krump, M. Simor, I. Hudec, M. Jasso, A. S. Luyt, *Appl. Surf. Sci.* **2005**, *240 (1-4)*, 268.
[4] S. Carlotti, A. Mas, *J. Appl. Polym. Sci.* **1998**, *69*, 2321.
[5] A. Raffaele-Addamo, E. Selli, R. Barni, C. Riccardi, F. Orsini, G. Poletti, L. Meda, M. R. Massafra, B. Marcandalli, *Appl. Surf. Sci.* **2006**, *252*, 2265.
[6] M. Prabaharan, N. Carneiro, *Indian J. Fibre Text.* **2005**, *252*, 2265.
[7] N. Carneiro, A.P. Souto, E. Silva, A. Marimba, B. Tena, H. Ferriera, V. Magalhàes, *Color. Technol.* **2001**, *252*, 2265.
[8] P. Yang, N. Huang, Y. X. Leng, Z. Q. Yao, H. F. Zhou, M. Maitz, Y. Leng, P. K. Chu, *Nucl. Instrum. Meth.* **2006**, *B 242*, 22.
[9] K. Chakrabarti, M Basu, S. Chaudhuri, A. K. Pal, H. Hanzawa, *Vacuum* **1999**, *53*, 405.
[10] H. U. Poll, U. Schladitz, S. Schreiter, *Surf. Coat. Technol.* **2001**, *142-144*, 489.
[11] F. Ferrero, *Polym. Test.* **2003**, *22*, 571.
[12] B. Gupta, J. Hilborn, CH. Hollenstein, C. J. G. Plummer, R. Houriet, N. J. Xanthopoulos, *Appl. Polym. Sci.* **2000**, *78*, 1083.
[13] M. M. Hossain, D. Hegemann, P. Chabrecek , A. S. Herrmann, *J. Appl. Polym. Sci.* **2006**, *102*, 1452.
[14] M. M. Hossain, A. S. Herrmann, D. Hegemann, *Plasma Process. Polym.* **2006**, *3*, 299.
[15] D. Hegemann, M. M. Hossain, *Plasma Process. Polym.* **2005**, *2*, 554.
[16] D. Hegemann, *Indian J. Fibre Text.* **2006**, *31*, 99.
[17] K. R. Kull, M. L. Steen, E. R. Fisher, *J. Membr. Sci.* **2005**, *246*, 203.
[18] T. R. Gengenbach, X. Xie, R. C. Chatelier, H. J. Griesser, *Polym. Prepr.* **1993**, *34*, 104.
[19] F. Arefi-Khonsari, M. Tatoulian, J. Kurdi, S. Ben-Rajeb, J. Amouroux, *J. Photopolym. Sci. Tec.* **1998**, *11*, 277.
[20] T. R. Gengenbach, R. C. Chatelier, H. J. Griesser, *Polym. Prepr.* **1997**, *38*, 1004.
[21] H. J. Griesser, L. Dai, T. R. Gengenbach, X. Xie, R. C. Chatelier, *Polym. Prepr.* **1997**, *38*, 1.
[22] N. Dilsiz, *J. Adhes. Sci. Technol.* **2000**, *14 (7)*, 975.
[23] D. Jocic, S. Vilchez, T. Topalovic, A. Navarro, P. Jovancic, M. R. Julia, P. Erra, *Carbohydrates Polym.* **2005,** *60*, 51.
[24] *"Industrial Dyes: Chemistry, Properties, Applications"*, K. Hunger, Eds., Wiley-VCH, Weinheim **2003**.
[25] D. Hegemann, A. Fischer, *J. Vac. Sci. Technol. A* **2005**, *23*, 5.
[26] D. Hegemann, H. Brunner, C. Oehr, *Nucl. Instrum. Meth. B* **2003**, *208*, 281.
[27] D. A. Waldman, Y. L. Zou, A. N. Netravali, *J. Adhes. Sci. Technol.* **1995**, *9 (11)*, 1475.
[28] S. P. Louh, C. H. Wong, M. H. Hon, *Thin Solid Films* **2006**, *498*, 235.
[29] T. Mieno, T. Shoji, K. Kadota, *Jpn. J. Appl. Phys.* **1992**, *31*, 1879.
[30] A. Lazea, L. I. Kravets, B. Albu, C. Ghica, F. Dinescu, *Surf. Coat. Technol.* **2005**, *200*, 529.
[31] Y. L. Zou, A. N. Netravali, *J. Adhes. Sci. Technol.* **1995**, *9 (11)*, 1505.

[32] M. Keller, A. Ritter, P. Reimann, V. Thommen, A. Fischer, D. Hegemann, *Surf. Coat. Technol.* **2005**, *200*, 1045.
[33] K. Schröder, A. Meyer-Plath, D. Keller, W. Besch, G. Babucke, A. Ohl, *Contrib. Plasma Phys.* **2001**, *41 (6)*, 562.
[34] D. F. Franceschini, *Braz. J. Phys.* **2000**, *30 (3)*, 517.
[35] R. R. Deshmukh, N. V. Bhat, *Mat. Res. Innovat.* **2003**, *7*, 783.
[36] J. Bico, U. Thiele, D. Quéré, *A. Physicochemical Eng. Aspects* **2002**, *206*, 41.
[37] J. Bico, C. Tordeux, D. Quéré, *Europhys. Lett.* **2001**, *55 (2)*, 214.
[38] N. Inagaki, K. Narushim, N. Tuchida, K. Miyazaki, *J. Polym. Sci. Pol. Phys.* **2004**, *42*, 3727.
[39] M. Creatore, P. Favia, G. Tenuto, A. Valentini, R. d'Agostino, *Plasma Polym,* **2000**, *5 (3/4)*, 201.
[40] K. M. Siow, L. Britcher, S. Kumar, H. J. Griesser, *Plasma Process. Polym.* **2006**, *3*, 392.
[41] *"Luminous Chemical Vapor Deposition and Interface Engineering"*, H. Yasuda, Eds., Marcel Dekker, New York, **2005**.

9. Deposition of Permanent Superhydrophilic a-C:H:N Films on Textiles

9.1. Introduction

The hydrophilic modification of textile surfaces using a dry and environmentally clean plasma technology has great importance especially for hydrophobic textiles such as polyester, polypropylene, acetate etc. to obtain enhanced properties for technological applications.[1] Textiles based on synthetic fibers often reveal a hydrophobic nature due to the lack of polar functional groups. The hydrophobic nature of such fabrics limits their application areas; while tailoring the surface chemistry in the top-most fabric layers by plasma might be an attractive way to overcome these complications. However, the heterogeneous complex structure of textiles has to be considered when applying plasma technology. Moreover, the permanence of a hydrophilic treatment is another factor for industrial applications. In Chapter 3, it was shown that a good hydrophilization and wettability of textiles can be achieved by an activation of polymer surfaces using non-polymer forming plasmas.[1-3] The hydrophilization was not durable for longer time periods (approx. 2 weeks) due to re-organization processes on the treated surface, which is one major disadvantage of such activation processes.[1] Kiyoharu et al.[4] reported that superhydrophilicity with water contact angles (CAs) below 10° and superhydrophobicity (CAs>150°) can be achieved by heat-treatment or irradiation of UV-light. These processes show limitations for substrate materials such as textiles due to the high temperatures (350-500°C) required. For example, the mechanical properties are greatly affected at high temperatures due to degradation/damage of polymer chains.

Using plasma technology, the properties of nitrogenated amorphous hydrocarbon films (a-C:H:N films) can be changed radically from polymer-like soft coatings to diamond-like (DLC) hard coatings by varying the process conditions.[5] Soft coatings of a-C:H:N with the desired surface chemistry were recently developed that meet the technological and ecological needs of textiles and clothing regarding strong hydrophilicity/hydrophobicity,[6] bacterial adhesion, protein adsorption etc.[7-9] More recently, it was found that a-C:H:N coatings show some advantages for substrate independent textile dyeing due to their nanoporous structure.[10] Lazar and Lazar showed that the incorporation of nitrogen containing functional groups such as amines, amides, imines, cyanides etc. within a hydrocarbon film facilitates the formation of voids in the coating yielding soft, hydrophilic coatings at low bias voltage.[11] In addition, the surface roughness of the C-H-N system might increase surface wettability and hydrophilicity.[12] The resultant films retained strong adhesion to the substrate, while they were highly crosslinked, chemically and mechanically stable.

Therefore, it has been investigated the deposition of nanoporous functional coatings by RF plasma at low temperature (room temperature) in order to obtain a permanent hydrophilic surface modification. First, a surface activation step was performed in a non-polymer forming plasma[13-16] and subsequently, nitrogen-containing nanoporous hydrogenated carbon films (a-C:H films) were deposited at low bias voltage (<30 V) with ammonia/hydrocarbon gaseous mixtures. The formation of crosslinked a-C:H:N

coatings with incorporated polar functionalities and a suitable texturing of the surface was simultaneously optimized. This hydrophilic surface treatment was found to overcome problems related to the conventional surfactants which had been previously developed.[17-19] Hence, this approach was found to maintain the hydrophilic properties at extended storage times and a higher mechanical stability than the aforementioned plasma activation followed by surfactant-coating.
The chemical composition of the a-C:H:N films was characterized by XPS. Dyeing of the deposited films provided information about film properties such as accessible functionalities within the coating, coating porosity, uniformity and purity etc. Surface wetting and hydrophilicity were characterized by CA measurements. AFM was used for a morphological characterization of the deposited nano-films. The thermal stability of the coating was verified by wash fastness tests according to an ISO standard (ISO 105 A02/A03). The mechanical and structural stability of the coatings and fabric surface wear, specifically abrasion, were characterized using an Abrasion & Pilling Tester and a crock meter.

9.2. Experimental

9.2.1. Materials and Plasma Treatments

The substrates and gases used in this study were described in Chapter 7. The pilot-scale web-coater and its vacuum system are described in detail in Chapter 6.[10,20] A two-step plasma process was carried out as described in the previous chapter.

9.2.2. XPS Analysis and AFM Measurements

XPS analysis and AFM measurements were performed according to the conditions described in Chapter 6 and Chapter 7, respectively.

9.2.3. Dyeing of a-C:H:N Films

The light-shade dyeing was performed according to the conditions described in Chapter 6. Since the PES substrate contains no cationic groups, an untreated substrate remained un-dyed (i.e. the dye does not interact with the substrate). Thus, the chemistry is provided by the coating and the dyeing is independent of the substrate material. The dyeing mechanism (film/dye interaction) was described in Chapter 8.[21]

9.3. Result and Discussion

9.3.1. Nitrogen Incorporation and Film Growth

The fragmentation of acetylene ($HC{\equiv}CH$) or ethylene ($H_2C{=}CH_2$) in a glow discharge initiates the plasma polymerization. This involves mainly two processes: the detachment of hydrogen from the monomer and the bond scission or bond

opening of the unsaturated triple or double bond yielding mono- and divalent radicals, respectively, as described in the previous chapter.[10,20]
Figure 1 shows the deposition rate and nitrogen concentration of a-C:H:N films in the NH_3/C_2H_4 atmosphere as determined by XPS. The deposition rate was found to decrease with the increase of ammonia added to the monomer flow. On the other hand, nitrogen incorporation increased with NH_3/C_2H_4 ratio. A high number of nitrogen functionalities at a high deposition rate can be obtained by optimizing the gas ratio. A gas ratio of NH_3/C_2H_4 around 0.71-1.0 and NH_3/C_2H_2 around 1.0-1.25 was found to yield the optimum regarding amine functionalities. Likewise, higher ammonia flow in the monomer gases C_2H_4 and C_2H_2 indicates a decrease in deposition rate; at the same time nitrogen content increases, as can be seen in Figure 1 and 2. Moreover, NH_3/C_2H_2 plasma yields a lower nitrogen content compared to NH_3/C_2H_4 plasma owing to the higher carbon content, which can be seen for equal gas ratios (ammonia/monomer = 1.25). These results are consistent with the deposition rates for both plasmas (NH_3/C_2H_2 or NH_3/C_2H_4). The previous study also confirmed the chemical composition of the a-C:H:N deposited films discussing the XPS spectra (C1s, N1s and O1s) elaborately.[22]

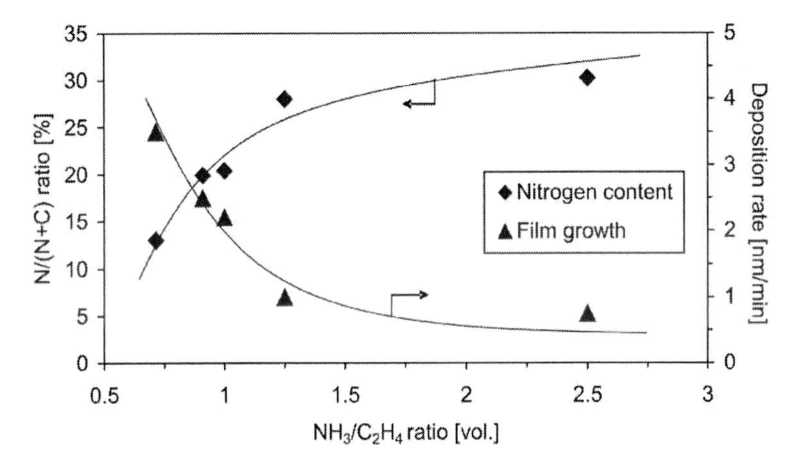

Figure 1. Nitrogen concentration and deposition rate at constant power (600 W) and different gas ratios (NH_3/C_2H_4).

9.3.2. Surface Porosity and Hydrophilicity

The surface roughness of plasma (NH_3/C_2H_2) coatings is summarized in Table 1, as calculated from AFM images of 500x500 nm^2 area. This analysis revealed that the surface roughness of the deposited films was not altered significantly by varying gas ratios of NH_3/C_2H_2 at a suitable, moderate power input (475 W). At higher power levels the surface roughness can be increased as described by Park et al.,[23] while precursor species predetermine the plasma polymerization rate, a variation of the gas ratio causes significant changes in film properties, in particular surface porosity. Figure 3 shows that the surface generated at higher NH_3/C_2H_2 ratios exhibits larger pores (<50 nm as obtained from AFM images), i.e. a distinct texturing, compared to the surface deposited at a lower NH_3/C_2H_2 ratio (<25 nm) as shown in Figure 4. An analysis of the surface area ratio (the ratio between porous surface and plane surface) also confirmed these findings. At higher NH_3/C_2H_2 ratios, the surface area ratio was found to be higher than at lower NH_3/C_2H_2 ratios. This means that an increased ammonia/hydrocarbon ratio leads to a pronounced nanoscaled texturing of the surface, and a high number of functionalities, especially at the coating surface (see Figures 1-2 and Table 2) leading to a high surface area ratio (Table 1). For coatings deposited from NH_3/C_2H_4 plasmas AFM measurements showed comparable results regarding texturing and roughness. These conditions resulted in the deposition of a homogeneous coating (Figure 3).

Table 1. Surface roughness of nanoporous a-C:H:N films (475 W).

Parameters	$NH_3/C_2H_2 = 4.0$	$NH_3/C_2H_2 = 1.25$
Average roughness R_{avg}	0.13 nm	0.14 nm
Root mean square roughness R_{rms}	0.16 nm	0.18 nm
Peak to peak roughness R_{pp}	1.20 nm	1.46 nm
Surface area ratio	0.4	0.2

Table 2. Depth profiling and elemental composition of a-C:H:N films. The power input was 475 W for C_2H_2 and 500 W for C_2H_4.

Gas ratio (vol.)	Without sputtering			After 30 s sputtering		
	C1s%	N1s%	O1s%	C1s%	N1s%	O1s%
$NH_3/C_2H_2 = 2.70$	63.5	22.6	15.6	78.7	19.5	1.7
$NH_3/C_2H_4 = 1.70$	65.0	22.5	12.5	81.6	14.7	3.6

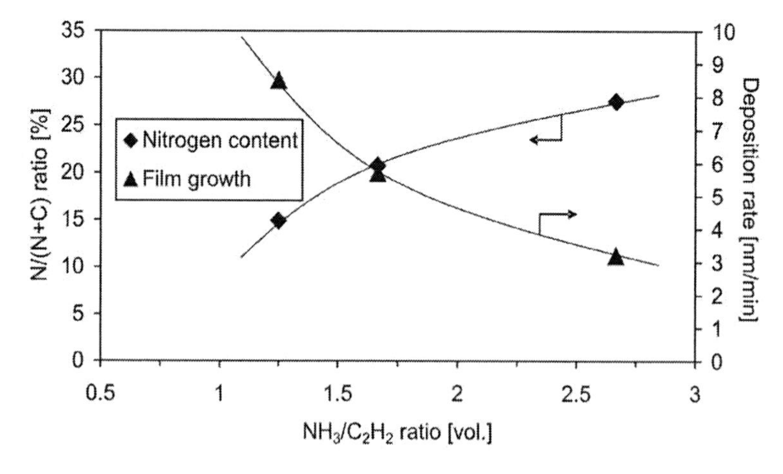

Figure 2. Nitrogen concentration and deposition rate at constant power (475 W) and different gas ratios (NH_3/C_2H_2).

Surface hydrophilicity was greatly enhanced due to the addition of polar functional groups contributing to a high surface porosity and imparting a textured surface, which yielded superhydrophilicity (water CA <10°)[4] because of the following two reasons. First, the attraction of polar functionalities in the coating expedites liquid (e.g. water) adsorption onto the surface and then, liquid interacts with the porous assemblies by wicking and diffusion processes and as a consequence, liquid transport into the coating with capillary spaces. Second, liquid drop follows surface morphology by capillary forces and then, the liquid is absorbed into/onto the porous and textured surface. Consequently, the liquid spreading and imbibition processes lead to a strong reduction in CAs (i.e. superhydrophilic surface) as can be seen in Figure 5.[24,25] Note that this effect is not due to swelling of the coating, as is known to occur with hydrogels, because superhydrophilicity for such thin coatings (2-3 nm) on flat surfaces was observed. Aging in ambient atmosphere indicated a good permanent hydrophilicity, while conventional plasma-activated surfaces are more prone to aging effects (Figure 5).
Interestingly, the peak to peak roughness R_{pp} is enhanced for lower NH_3/C_2H_2 ratios. The inner pore structure might become more pronounced at lower NH_3/C_2H_2 ratios and the peaks may become thicker and valley may form resulting from higher R_{pp} roughness (in some cases peak to valley roughness) (see Table 1 and Figure 4). Moreover, it can be assumed that less radical sites are left at the surface after plasma polymerization, yielding less oxygen functional groups by post-plasma reactions. Therefore, lower hydrophilicity and wettability were observed at these lower NH_3/C_2H_2 ratios due to a less suitable texturing and fewer polar groups at the surface (Figure 5). Nevertheless, these coatings revealed almost no aging effects and showed improved dyeability (see Chapter 8).

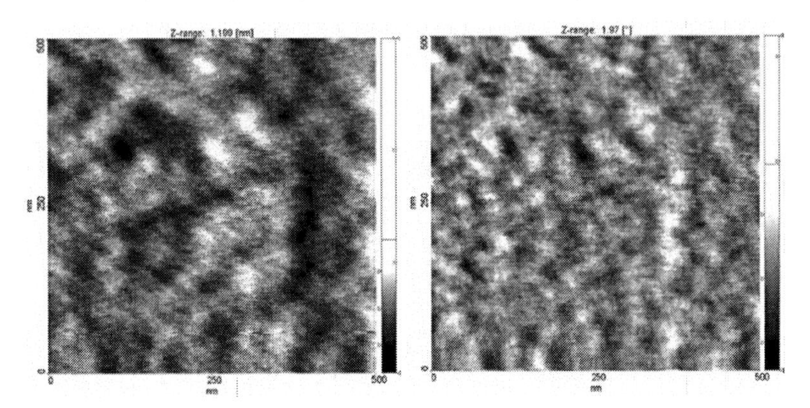

Figure 3. Surface topography of a-C:H:N films (500 W, NH_3/C_2H_2 = 4.00). Left: surface topography, right: phase image.

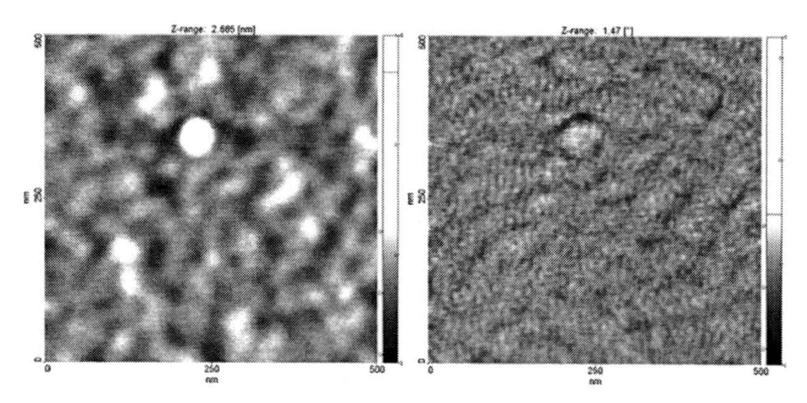

Figure 4. Surface topography of a-C:H:N films (500 W, NH_3/C_2H_2 = 1.25). Left: surface topography, right: phase image.

Hence, to obtain permanent, superhydrophilic coatings on textiles leading to complete wetting, plasma operating conditions including plasma power and gas ratio should be chosen in careful consideration regarding the aspects discussed, in order to obtain rivaling deposition/etching processes, while the gas pressure was kept constant (at 10 Pa).

9.3.3. Contact Angle and Surface Hydrophilicity

Static CAs were measured on fabrics and glass substrates with de-ionized water in a conditioned room (65% RH, 20°C) via optical methods[10] with a drop size of ≈ 10 μL. Static CAs on treated fabrics could not be measured due to complete wetting, as the water drop was observed to disappear quickly, whereas control PES fabrics display their hydrophobic nature with a CA of $\approx 80°$. Storage of plasma-coated textile fabrics for more than one year under ambient conditions proved the longevity of the superhydrophilic properties assuring complete wetting over time. The coated glass substrates facilitated the observation of minor differences in CAs with respect to aging. Figure 5 shows a comparison of static CAs at different plasma treatments and corona treatment. Higher CAs and distinct aging effects were found for the corona treatment and Ar/O_2 plasma activation, which was performed on smooth PET foils. A lower number of polar functionalities, less crosslinking and more damaging of the modified surfaces resulted in higher CAs. The rapid change in CAs with aging observed for these plasmas can be ascribed to restructuring of the polar functional groups.[26] On the contrary, hydrophilicity is evidently improved using ammonia/hydrocarbon gaseous mixtures and deposition of a-C:H:N plasma coatings as compared to the other two plasma types. The remarkable reduction in aging indicates highly crosslinked plasma polymers thanks to the strong hydrocarbon network. Higher NH_3/C_2H_2 ratios were found to yield the lowest CAs due to the formation of polar functionalities on the plasma polymer surface. Free radicals created on the plasma polymer produced polar oxygen functionalities, when the treated substrates were exposed to atmosphere, although no oxygen was used during plasma polymerization. Similar results were obtained for NH_3/C_2H_4 gaseous plasmas.[10] The compositional analysis showed that oxygen functionalities such as carbonyl, carboxyl etc. formed mostly on the coated surface, as after 30 s sputtering the oxygen content was strongly reduced (Table 2).

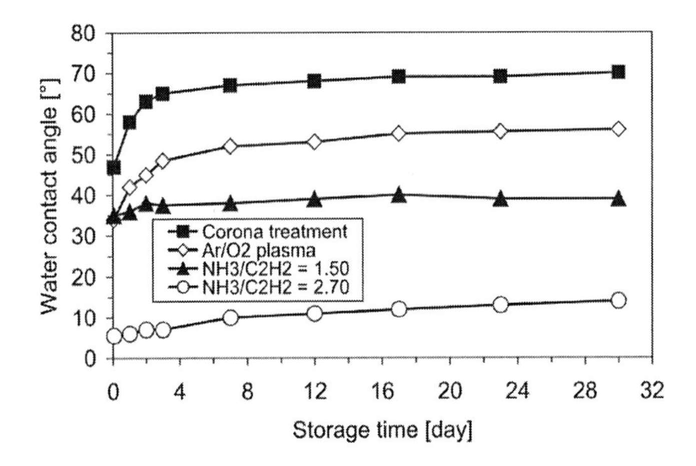

Figure 5. Obtained water contact angles measured for different plasma treatments on smooth PET foils and coated glass substrates depending on storage time. Untreated PET foils show their hydrophobic nature with a CA of 75-80°.

9.3.4. Dyeing and Coating Quality

Dye molecules (a few nanometers in size) are able to penetrate into nanoporous films and facilitate the dyeing of a-C:H:N thin films by forming chemical bonds mainly with amino groups[10,27] within the plasma coatings. Since the dyestuff used specifically binds with amine groups, dyeing can be used as a specific chemical tracer to detect the amount of amine groups inside the coatings. Table 3 shows color values as a function of plasma power, film thickness (<100 nm), and nitrogen content (by XPS) within the plasma coatings. Plasma power in the range of 500-1000 W at constant gas ratios was found not to influence the obtained nitrogen content, while the deposition rate varied strongly depending on the power input due to an increase in monomer fragmentation, which is decisive for film growth. The deposition rate increased, depending on the gas ratio, as the power increased, with nitrogen content remaining constant (20%). These results are consistent with Balazs et al.[7] It is assumed that the same film forming radicals are obtained for a fixed gas ratio over the examined power input range, thus not increasing N/C ratio. However, increasing power input yields more interaction during film growth and thus different functionality of the film. On the other hand, the deposition rate was found to drop again at very high power levels, i.e. 1000 W, due to etching and sputtering effects.[28] It was observed that the color intensity is dependent on the amine group concentration within the coating, thus scaling linearly with film thickness[29] for a fixed set of plasma parameters. These observations might be explained as follows: higher NH_3/C_2H_2 ratios (i.e. high nitrogen content) produce more amides, cyanides, imines etc. within the coatings rather than simple amine functional groups. The highest nitrogen

content (30%) reveals the lowest color difference (DE^* = ≈26) due to a lower number of amine groups (Table 3).[22] Highly reactive ammonia, on the other hand, generates a high number of surface free radicals at higher nitrogen content (i.e high NH_3/C_2H_2 = 2.7) as compared to a lower ratio (NH_3/C_2H_2 = 1.7). These reactive radicals contribute to form oxygen-containing polar functional groups by post-plasma reactions in the atmosphere (Table 2). As a consequence, superhydrophilic coatings can be achieved with high ammonia content by incorporation of polar functional groups into the surface within a crosslinked hydrocarbon network. Low ammonia to monomer ratios, on the other hand, effectively lead to an enhanced dyeability by a higher incorporation of amine functional groups. High monomer flow produced high hydrocarbon formation and the creation of fewer free radicals. This resulted in the formation of a reduced number of oxygen functionalities on the surface and consequently, water CAs were found to be higher (Figure 5).
It is interesting to note that even as film thickness increased, the color intensity decreased, as power input increased at constant nitrogen incorporation (20%). This finding also indicates the presence of unsaturated stochiometry (for example, unsaturated bonds such as C=C, C=N etc.) which decline the color intensity, even when they have higher film thicknesses. Thus, chroma (C^* - brightness) comprise another important factor which should be considered when evaluating amine functionalities and coating quality with respect to coating functionality. At a constant nitrogen content of 20%, the hue (h°) of the color also decreased as the power input increased. This is essentially due to the reduction in color purity, i.e. due to a decrease in blueness and brightness. Unsaturated bonds in the coating mainly cause color impurity (high yellowness). The depletion of the blue color (b^*) component was compensated for as the yellowness in the color increased. This result demonstrates that high plasma power levels (>600 W) produced more monomer gas fragmentation by hydrogen abstraction processes, forming unsaturated bonds in the plasma polymers. This effect is more pronounced for NH_3/C_2H_2-derived plasma coatings due to the triple bond of acetylene.

Table 3. CIELAB color values of dyed a-C:H:N films at different plasma power, film thickness, and nitrogen concentration (NH_3/C_2H_4 plasma).

Power input (watt)	Film thickness (nm)	N-content in% [N/(N+C)]	CIELAB color values (D65/10°)					
			DE^*	C^*	h°	L^*	a^*	b^*
700	52	13	33.0	17.6	225.9	71.4	-12.2	-12.6
800	58	13	31.5	16.4	223.1	72.1	-12.0	-11.2
600	54	20	43.0	21.6	232.1	65.9	-13.3	-17.0
900	70	20	39.5	19.3	228.2	65.2	-12.8	-14.4
1000	62	20	36.4	15.9	217.1	66.1	-12.7	-9.6
500	30	30	26.0	13.9	228.5	77.1	-9.4	-10.3
700	30	30	25.8	12.5	228.9	80.9	-8.3	-9.4

9.3.5. Stability and Permanency of Plasma Coatings

The limiting factors for longevity of a coating are mainly given by the crosslinking and internal re-organization processes. The loss in structural integrity due to abrasion, e.g. through use and care, may limit the wear life-time of the coatings. The latter affects the coating's functional performance. Abrasion resistance is measured by subjecting specimen to rubbing motion in the form of a geometric figure against a wool fabric counterpart. Structural changes in dyed plasma-coated samples were thus tested for abrasion resistance under a load of 12 kPa on a Martindale instrument (Test method, SN 198514). It was evident that no changes were detected in the coated surface after 60,000 rubbing cycles, since no damage of the dyed coating could be observed. The coating adhered well to the substrate surface due to the high interfacial adhesion between plasma polymers and the substrate surface.

Table 4. Wash (60°C) and rub fastness of a-C:H:N films deposited on PES fabrics.

Gas ratio (vol.)	Wash fastness			Rub fastness	
	Color change	Staining (PES)	Staining (wool)	Dry	Wet
$NH_3/C_2H_2 = 1.00$	3	5	4	4-5	4-5
$NH_3/C_2H_2 = 1.25$	3	5	4	4	4
$NH_3/C_2H_2 = 1.50$	3	5	4	4	4-5
$NH_3/C_2H_4 = 0.71$	3	5	3-4	4	4-5
$NH_3/C_2H_4 = 1.00$	3	5	3-4	4	4-5
$NH_3/C_2H_4 = 1.25$	3	5	3-4	4	4-5

The wash and rub fastness of the dyed samples were tested by the ISO test method. Under D65 illumination color changes, staining and rub were evaluated using grey scales: ISO-105-A02 grey scale for assessing change in color; ISO-105-A03 grey scale for staining and rub. An evaluation of wash fastness showed good to excellent results (Table 4). The fastness tests did not show any significant difference between the samples. Furthermore, the dyeing uniformity validated the coating homogeneity. Since nanoscaled coatings (<100 nm) were deposited, the touch and comfort of the textiles were not affected. Moreover, the air permeability of the coated textiles was maintained, while wet-chemical modifications may have affected the porosity of textiles.

9.4. Conclusion

The deposition of nitrogen containing superhydrophilic nanoporous a-C:H films was investigated using low pressure RF plasma at low bias voltage and NH_3/C_2H_2 and NH_3/C_2H_4 gaseous mixtures. The modification yielded a permanent change from a hydrophobic (CAs = $\approx$80°) to a superhydrophilic surface (complete wetting) by nanometer-thick plasma coatings on textile PES fabrics. Both plasmas contribute to film growth by forming mono-/divalent radicals by hydrogen detachment and bond scission, respectively. Ammonia to monomer gas ratios and plasma power levels were found to be very important in order to obtain nanoporous a-C:H:N films. A high amount of accessible amine-functional groups within the nanoporous coatings, as detected by dyeing with acid dye, could be obtained at suitable ammonia to hydrocarbon ratios (around 1.0), while they exhibited a moderate hydrophilicity. A more pronounced texturing of the plasma-coated surfaces due to a modified surface topography was obtained at higher NH_3/C_xH_y ratios. As a result, the liquid spreads at the porous film, facilitated by surface polar groups within a crosslinked hydrocarbon network. This leads to an important reduction in CAs and to reduced aging effects. Therefore, permanent superhydrophilic coatings can be obtained. High ammonia flow produced an increased number of reactive sites (such as free radicals) at the surface, while having a reduced deposition rate. As a result, oxygen polar functionalities, which are decisive for the formation of superhydrophilic surfaces, can be incorporated by post-plasma reactions.

A relatively high monomer flow contributed to the production of an increased number of amine functionalities in the coating showing higher dyeability. Thus, a dyeable, hydrophilic modification can be obtained by a suitable combination of gas ratio and film growth offering a high number of amines incorporation. It is very interesting to find that at constant nitrogen content levels, the color intensity was decreased at high plasma power due to unsaturated bonds increasing the yellowness index. Therefore, dyeing can be used as a measure of coating quality and uniformity.

The coating adhered well to the substrate and was structurally and mechanically stable. Plasma-coated PES fabrics were found to maintain their complete wetting behavior, and thus are able to support excellent adhesion to subsequent coatings or lamination for more than one year of storage at ambient conditions. These superhydrophilic plasma coatings overcome the traditional problems of high application temperature of conventional surfactants and heat treatment modification methods for textiles. Nitrogenated a-C:H coatings (thickness <100 nm) have a great potential due to their mechanical stability, elasticity and flexibility, while leaving the fabric comfort and touch unaffected.

9.5. References

[1] M. M. Hossain, D. Hegemann, P. Chabrecek, A. S. Herrmann, *J. Appl. Polym. Sci.* **2006**, *102*, 1452.

[2] C. W. Kan, K. Chan, C. W. M. Yuen, M. H. Miao, *J. Mat. Proc. Tech.* **1998**, *82*, 122.

[3] C. Canal, R. Molina, P. Erra, A. Ricard, *EUR. Phys. J.-Appl. Phys.* **2006**, *36*, 35.

[4] K. Tadanaga, J. Morinaga, T. Minami, *J. Sol-Gel Sci. Techn.* **2000**, *19*, 211.

[5] L. A. Romanko, A. G. Gontar, A. M. Kutsay, S. I. Khandozko, V. Yu. Gorokhov, *Diam. Relat. Mater.* **2000**, *9*, 801.

[6] K. R. Kull, M. L. Steen, E. R. Fisher, *J. Membrane Sci.* **2005**, *246*, 203.

[7] D. J. Balazs, M. M. Hossain, E. Brombacher, G. Fortunato, E. Körner, D. Hegemann, *Plasma Process. Polym.* **2007**, *4*, S380.

[8] M. Loughran, S. Tsai, K. Yokoyama, I. Karube, *Curr. Appl. Phys.* **2003**, *3*, 495.

[9] M. Taborelli, L. Eng, P. Descouts, J. P. Ranieri, R. Bellamkonda, P. Aebischer, *J. Biomed. Mater. Res.* **1995**, *29*, 707.

[10] M. M. Hossain, J. Müssig, A. S. Herrmann, D. Hegemann, *J. Appl. Polym. Sci.* **2009**, *111*, 2545.

[11] G. Lazar, I. Lazar, *J. Non-Cryst. Solids* **2003**, *331*, 70.

[12] P. Yang, N. Huang, Y. X. Leng, Z. Q. Yao, H. F. Zhou, M. Maitz, Y. Leng, P. K. Chu, *Nucl. Instrum. Meth. A* **2006**, *B 242*, 22.

[13] M. Keller, A. Ritter, P. Reimann, V. Thommen, A. Fischer, D. Hegemann, *Surf. Coat. Technol.* **2005**, *200*, 1045.

[14] N. Inagaki, *"Plasma Surface Modification and Plasma Polymerization"*, Technomic, USA, **1996**, pp. 28-29.

[15] G. Placinta, F. Arefi-Khonsari, M. Gheorghiu, J. Amouroux, G. Popa, *J. Appl. Polym. Sci.* **1997**, *66*, 1367.

[16] D. Hegemann, A. Fischer, *Proc. Int. Textile Congress,* Terrassa/Spain, October 18-20, **2004**.

[17] H. Y. Kim, B. H. Kang, *Appl. Therm. Eng.* **2003**, *23,* 449.

[18] J. S. Cho, Y. W. Baeg, S. Han, K. H. Kim, J. Cho, S. K. Koh, *Surf. Coat. Technol.* **2000**, *128*, 66.

[19] S. C. Ha, C. H. Kim, S. P. Ahn, G. A. Dreitser, Proc. Int. Conf. Exhibition, Lisbon, Portugal, **1998**, pp. 423.

[20] D. Hegemann, *Indian J. Fibre Text.* **2006**, *31*, 99.

[21] D. Jocic, S. Vilchez, T. Topalovic, A. Navarro, P. Jovancic, M. R. Julia, P. Erra, *Carbohyd. Polym.* **2005,** *60*, 51.

[22] M. M. Hossain, D. Hegemann, A. S. Herrmann, *Plasma Process. Polym.* **2007**, *4*, 1068.

[23] Y. Park, S. Rhee, *Surf. Coat. Technol.* **2004**, *179*, 229.

[24] J. Bico, U. Thiele, D. Quéré, *A. Physicochemical Eng. Aspects* **2002**, *206*, 41.

[25] J. Bico, C. Tordeux, D. Quéré, *Europhys. Lett.* **2001**, *55 (2*), 214.

[26] M. M. Hossain, A. S. Herrmann, D. Hegemann, *Plasma Process. Polym.* **2006**, *3*, 299.

[27] K. M. Siow, L. Britcher, S. Kumar, H. J. Griesser, *Plasma Process. Polym.* **2006**, *3*, 392.

[28] D. Hegemann, M. M. Hossain, *Plasma Process. Polym.* **2005**, *2*, 554.

[29] D. Hegemann, M. M. Hossain, D. J. Balazs, *Prog. Org. Coat.* **2007**, *58*, 237.

10. Nanoporous Coatings for Multifunctional Applications

10.1. Nanoporous Coatings for UV Protecting Absorbers

Textile research has been paying attention to fabric protection against ultraviolet (UV) radiation only in the last two decades. Intense UV radiation can result skin damage such as sunburn, allergies, and even skin cancer.[1,2] Textile can provide effective protection against such damages. For outdoor activities, the best way to protect the skin against UV radiation is clothing. In addition, many textiles such as aramid etc. are very sensitive to UV, they lose their strength under UV radiation due to fiber degradation. Therefore, the UV protection of textiles against photodestruction is of high practical interest.

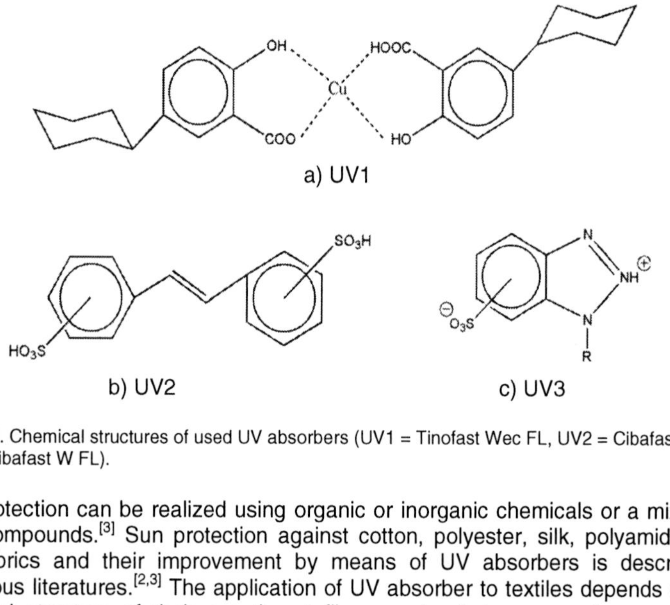

a) UV1

b) UV2

c) UV3

Figure 1. Chemical structures of used UV absorbers (UV1 = Tinofast Wec FL, UV2 = Cibafast N3 and UV3 = Cibafast W FL).

The protection can be realized using organic or inorganic chemicals or a mixture of both compounds.[3] Sun protection against cotton, polyester, silk, polyamide, wool etc. fabrics and their improvement by means of UV absorbers is described in numerous literatures.[2,3] The application of UV absorber to textiles depends on their chemical structure of their constituent fibers and substances such as additives, textile processing aids etc. present in and on the textiles.[3-7] Therefore, a specific reactive UV absorber and its application condition are limited generally to each fiber type. Moreover, if the fabrics are closely woven or knitted, their UV protection performance can be improved, on the other hand the porosity, i.e. air permeability, of the fabrics will be reduced which is also a disadvantage for clothing comfort. In this case, nanoporous coated textiles can be used to incorporate UV absorbers. Thus, the amounts of absorbed UV absorber on textiles can be increased without using additional chemicals or additives.

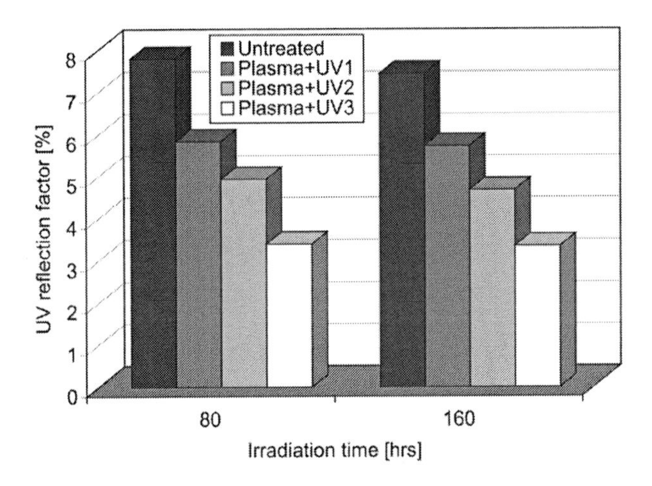

Figure 2. Nanoporous coatings improved UV protection performance on aramid fabrics depending on the type of UV absorbers.

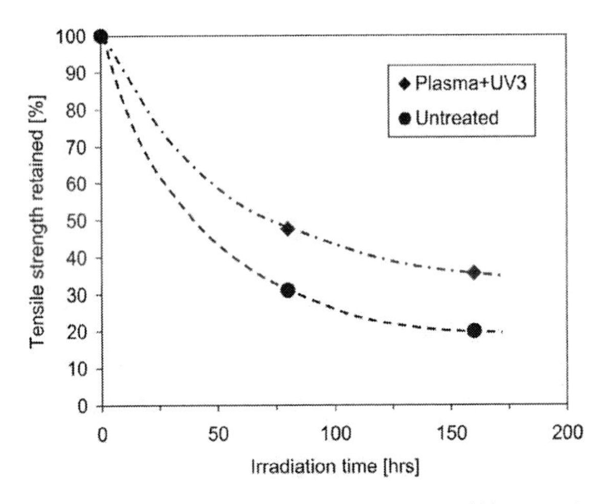

Figure 3. Mechanical stability of aramid fabrics against UV exposure within an accelerated aging test (plasma coating thickness ≈100 nm).

The UV absorbers, namely Tinofast Wec FL (UV1), Cibafast N3 (UV2) and Cibafast W FL (UV3), used in this work were supplied by Ciba-Geigy Limited (Switzerland). The chemical structures of the absorbers are shown in Figure 1. UV absorbers were applied to plasma-coated aramid fabrics. The entire nanoporous coating (≈100 nm) contains amines, carboxyl groups etc., which can chemically bind with the functional groups of UV absorbers. Thus, a permanent UV absorber-fiber bond can be obtained. Figure 2 shows UV reflection (EN 410) on aramid fabrics as a function of irradiation time. In all cases, plasma-coated fabrics show improved UV protection as compared to untreated fabrics. This can be explained by the screening action of the

UVA in the UV spectral range and thus decreasing the rate of photochemical reactions. UV3 found to be very efficient as compared to the other two types. This is because UV3 has two functional groups (-SO_3 and -NH) in its structure. They can form salt linkage with the amine or carboxyl containing plasma coating. Therefore, a high amount of UV3 can be attached to plasma coated fabrics and thus yielding high UV absorber-uptake to aramid fabrics. On the other hand, since UV2 has only -SO_3 functionalities, which has one possibility to form ionic bonds with amine groups in the coating, but it has no interaction with –COO in the coating. Thus, UV2 exhibited less performance against UV radiation. Furthermore, the metal containing UV1 compound was found to be rather difficult to attach to plasma coating due to their complex structure. As a consequence, the amount of incorporated UV1 could not very effectively prevent UV radiation reactions. Moreover, the strength decrease by UV radiation could be dramatically reduced by using UV3 compared to untreated sample (Figure 3).

Self-cleaning Surface

Incorporating fluorocarbon (FC) monomers into the nanoporous coatings, they can provide water repellent properties with a low surface energy. Since FC groups are known to impart hydrophobic behavior, superhydrophobic surface on wettable plasma-coated PES textiles was obtained, showing a CA greater than 140° (Figure 4). Additionally, the coatings are well resistance against oil (oil grade 6, AATCC 118). Therefore, the FC monomer-loaded a-C:H:N plasma coatings can be utilized as self-cleaning (stain-repellent) surfaces. Due to the permanent bonding between the FC monomers and plasma polymers, the chemically modified plasma-polymer composites have very strong mechanical stability against rubbing and wear which was examined by a Martindale Test (Test method, SN 198514), as can be seen from Figure 4. The CAs were found to be 140° and more even after 80,000 rubbing cycles. This indicated that the coatings adhere well to the substrate demonstrating high mechanical performance.

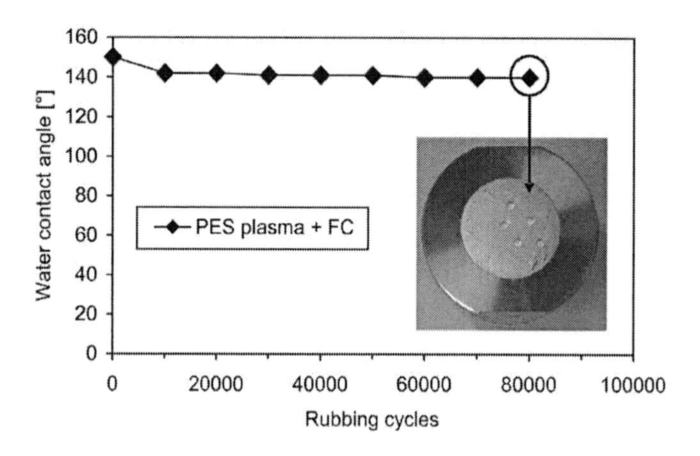

Figure 4. Mechanical stability and superhydrophobicity of PES textiles (Martindale Test).

10.2. Nanoporous Coatings and Fiber-reinforced Composites

Textile fibers, both natural and synthetic, have found increasing use as reinforcing elements in fiber-reinforced composites due to their low density combined with exceptional mechanical properties.[8] Synthetic fibers such as glass, carbon etc. are mostly used fibers providing strength and durability for engineering thermosetting and thermoplastic resins to make composites. Because of their high strength and temperature resistance, these rather expensive fibers are a particular choice for high performance applications such as in aerospace. A major disadvantage of these fibers is their non-biodegradability. Natural fibers are abundant with a low-volumetric cost, renewable, light weight, recyclable and biodegradable, which causes their ever-increasing interest in reinforced composites compared to synthetic fibers.[9,10] J. Müssig reported the perspective use of natural fibers in composites on the basis of sustainability for the future – economy, ecology and society.[11] The combination of interesting mechanical and chemical properties together with their ecological and environmentally friendly character has triggered the worldwide demand for "green composites". For example, natural fiber reinforced composites are of particular interest for the European automotive industry. Daimler-Chrysler research group has found that flax and hemp fibers are potential candidates to replace glass fibers in environmentally safety engineering composites such as door panels etc.[12]

The mechanical properties of fiber reinforced composites mainly depend on:[13-15]

- Fiber and matrix properties
- Physical and chemical characteristics of the fiber surface and
- Mechanism of load transfer from fiber to matrix at the interface.

As a result, main attention is usually given to the surface texture and properties of these fibers. The adhesion between the treated fiber and the matrix does play a major role in determining interfacial strength in composites. Surface properties of fibers can be improved, widely used, by various methods such as mechanical, chemical, or plasma techniques.[13,14,16] Wetting agents, sizings (coatings), coupling agents such as silane, isocynate etc. based compounds can be used chemically to improve bonding between fibers and coupling agent (usually by covalent bonds) or between coupling agent and matrix materials (secondary weak bonds), as described in numerous literatures.[17-19] However, chemical modification may have some disadvantages, such as corrosion problem for oxidation of fibers by nitric acid, decrease in fiber strength, environmental pollution etc. In this regard, plasma technique is becoming increasingly popular because it is an applicable and dry process causing no pollution.[20] The main purpose of surface modification of fibers used as reinforcements in composite materials is to modify the chemical and physical structure of the surface layers, tailoring fiber-matrix bond strength.

The nanoporous coatings as described in proceeding chapters have several advantages for fiber reinforced composites as compared to other processes. The coating exhibits the following advantages while keeping the bulk fiber properties unaffected.

- High surface area ensures enlargement of contact area and textured surface promotes mechanical interlocking. Thus, interfacial adhesion and strength between fiber and matrix can be improved.[21]
- Plasma coated fiber surface contains huge numbers of functional groups such as amines, carboxyl, hydroxyl etc. which can chemically interact with matrix materials. As a consequence, strong covalent bond can be obtained between fiber and matrix.[14]

Environmental friendly composite materials were prepared from biodegradable poly(lactic acid) (PLA)[22,23] with flax fibers (Holstein Flachs GmbH, Mielsdorf, Germany), where nanoporous coatings had been deposited on the flax fibers in order to incorporate additional functional groups on the fiber surfaces. The fiber-reinforced/PLA composites were prepared as described in the literature.[24] Briefly, PLA sheets were compression molded by two iron plates in a laboratory press unit (Rucks Maschinenbau GmbH, Glauchau, Germany). The nominal pressure was kept at 1000 kN for 10 min at temperature 217°C. Then, both fiber ends were glued carefully across the PLA sheet in order to preserve the fiber alignment. A second PLA sheet was placed over the first PLA sheet with the fibers. The specimen was pressed with 20 bars for 90 seconds.
To evaluate the effect of the modification, the single fiber fragmentation test was carried out in a tensile testing machine (Instron No. 4502H3358). 30 mm gauge length was used and the number of fiber breaks of the composite within the gauge length was counted. The fragment length of the fragmented specimen was measured using an optical Zeiss MC80 microscope. Critical fiber length "the shortest fragment length" was estimated from the average fragment length, i.e. the critical fragment length is equal to 4/3 of the average fragment length.
As can be seen in Table 1, the results showed that the interface properties of flax/PLA composites are quite promising for nanoporous coated fibers as compared to untreated fibers. A gas ratio of NH_3/C_2H_4 around 0.7-1.0 and NH_3/C_2H_2 around 1.0-1.25 and a discharge power of around 500-600 W were found to yield the optimum regarding amine functionalities in the nanoporous coatings and thus, resulting in obtaining high number fragments. A less number of fragments, i.e. poor fiber-matrix adhesion, were observed for optimized hydrophilic plasma-treated surfaces ($NH_3/C_2H_4>1.7$ and $NH_3/C_2H_2>2.7$). It should be mentioned that no fiber broken ends were found with untreated-fiber/PLA matrix at all indicating sliding of fibers within the composite due to low adhesion. Thus, the interfacial surface properties of composites were enhanced due to nanoporous coatings, yielding high fiber/matrix interfacial adhesion. The increased interfacial adhesion in fiber/PLA matrix may arise from structural interaction and from chemical reactions between modified fiber and PLA matrix such as H-bonding, dipole-dipole interaction, covalent bonding etc. Consequently, this finding provides some tentative evidence that the deposited coatings have high potential to enhance bonding between fiber and PLA matrix.

Table 1. Number of fragments, average fragment length and critical fiber length obtained from single fiber fragmentation tests for nanoporous coated fiber (gauge length 30 mm, load 0.2 mm/min).

Composite	No of fragments (mean)	Avg. fragment length (mm)	Critical fiber length (mm)
Flax – PLA	5.0	8.3	11.1

Untreated fibers are very hygroscopic and they naturally contain approximate 10% water in their structure (moisture absorption that causes swelling of the fibers),[16] which is a major problem for adhesion with polar PLA. In this regard, nanoporous coatings have been deposited under vacuum in a two step process, thus removing water without affecting the mechanical properties of fibers. The moisture uptake of the natural fibers can be minimized by plasma treatments. Pretreatment with Ar/O_2 plasma facilitated removing of pectic substances, wax etc. from fiber surface and functionalization of the fiber-surfaces by adding polar groups.[25] Subsequent nanoporous coatings preserved to shield –OH groups of fibers and at the same time, other functional groups such as amines, carboxyl, and hydroxyl groups etc. were incorporated on the fiber surfaces.[26] Thus, PLA can bind chemically to the fiber surfaces and thereby improving interfacial property between the fibers and the polymer matrix and hence improved composite properties.

10.3. Sputter Deposition

Sputter deposition is a method to deposit coatings on substrate by sputtering e.g. a metal target. Sputtering usually performed in a vacuum chamber using argon plasma. In magnetron sputtering, a magnetic field is applied parallel to the target surface. In this work, pulsed dc magnetron sputtering was used in order to prevent the formation of arcs which can damage the power supply or can cause uneven films. Arc-free operation requires the "on time" during pulsing to be sufficiently short to avoid charge build-up that cause breakdown and arching and the "off time" to be sufficiently long to fully discharge the surfaces in order to avoid charge accumulation.[27]

Sputtering occurs at the target (e.g. Cu) when an energetic ion (e.g. Ar^+) or atom makes a series of collisions with atoms in the target. One or all of the following phenomena may occur:

- The incident ion becomes a neutral before impact; i.e. it picks up an electron from the target, so that it is an energetic Ar atom that shares its momentum with Cu atoms. A Cu atom that gains enough energy from the collisions will be displaced from its normal lattice site and hence it is ejected from the target.[28] Thus, vapor atoms are continuously available to deposit on surfaces, at the same time, deposited atoms migrate on the surface and interact with each other as well as with the substrate atoms yielding "physisorption or chemisorption". The interactions determine the morphology of the film growth.
- The ion impact may cause the target to eject an electron, usually referred to as a secondary electron or may become implanted in the target "ion implantation".[29]
- Structural changes at the surface layer (target) can occur due to the ion impact and resulting collisions.

10.4. Orientation of Natural Fibers in Composites

The use of natural fibers for example in fiber-reinforced polymers requires fibers with very specific properties.[30] Fiber or bundle fineness improves the tensile properties of composites. Together with fiber length the fineness is a critical property for the quality of injection molded parts. [31] Fiber orientation is another important material property, since it plays a key role for the improvement of the mechanical stability of fiber reinforced composites. Furthermore, it is also necessary to measure fiber geometry and orientation inside the composite. Scanning techniques are a very cost efficient and reliable method to carry out these controls. [32] According to Walther et al.[33] wood fiber geometry and orientation is measurable in a Medium Density Fiberboard (MDF) using X-ray Microtomography (µCT). Metallized fibers can be used in order to observe these controls more precisely.

A pilot-plant plasma reactor was used to metallize hemp fibers, which is able to treat textiles semi-continuously with a maximum width of 65 cm. The sputter deposition was carried out with a pulsed dc magnetron sputtering (1 kW, 1 Pa). The fibers were coated twice by turning them in order to deposit approx. 125 nm thick Cu films. X-ray μCT was performed on composite samples with a dimension of 2X2X5 mm^3 at beam line BW2 at HASYLAB, Hamburg, Germany. The synchrotron radiation X-ray μCT-equipment was used. The photon energy for the scans was set to 10 keV to match the low absorption of the fibers.

The alignment and the distribution of the fibers in the sample are displayed in the cross sections shown in Figure 5. Using Cu-coated fibers the contrast of the fibers can be improved noticeably and thus, fiber orientation can be observed precisely (Figure 5, right). Further research is needed for the observation of microstructure (such as fiber size, fibrillation etc.) of natural fiber in composites, especially fiber orientation in matrix.

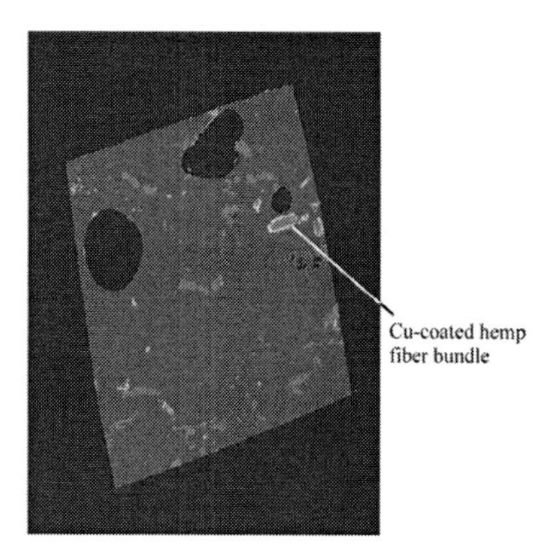

Figure 5. Uncoated hemp fibers in an Epoxy thermoset resin (left) and Cu-coated hemp fibers in an Epoxy thermoset resin (ca. 15% of the embedded hemp fibers were preliminarily sputtered with Cu) (cross sections: 2.90 X 2.90 mm^2 with pixel edge length 1.89 μm and spatial resolution 2.96 μm).

10.5. Silver Incorporated Nanoporous Composites

The antibacterial properties of silver (Ag), as well as copper and other metals, have been used for centuries in many technological applications. Since Ag exhibits the highest toxicity for microorganisms and is the least toxic to animal cells,[34,35] it has been used in recent years in various medical applications such as antimicrobials for wound dressing, reducing bacterial infections in medical devices etc. In the past ten years, the use of Ag as an antimicrobial agent in biomedical devices has been increased, especially in the fabrication of various catheters.[34] Furthermore, Ag can be potentially used for biomaterial surface enhancements and for electrical conductivity. Thus, Ag incorporated a-C:H:N composites can potentially produce multifunctional textiles surfaces by a combination of different properties due to the plasma polymer matrix and the Ag content, as shown in Figure 6. The wettability, bioactivity and functional group density, for example, can be tailored by the plasma polymer matrix mostly independent from the amount of Ag in the coatings. On the other hand, conductivity and antimicrobial properties increase with the increase of Ag content. In addition, mechanical stability of the Ag/a-C:H:N film decreases a little with the increase of Ag content (see Figure 6).

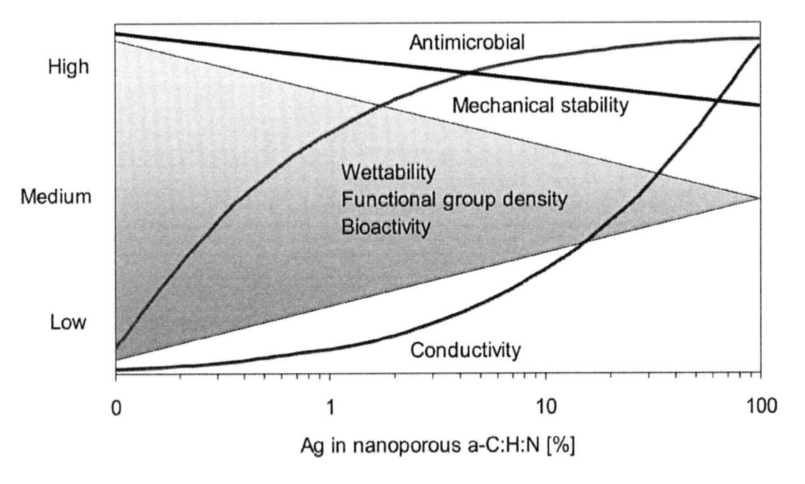

Figure 6. Different surface functionalities depending on the amount of Ag incorporation in a-C:H:N composites.

Plasma is one of the most effective means to incorporate Ag into organic films. Ag nanoparticle are incorporated in nanoporous a-C:H:N films using RF plasma polymerization/co-sputtering techniques.[36,37] There is a complex process of film growth occurring when a gas mixture of C_2H_4/Ar/NH_3 is used in the plasma discharge.[38] Two concurrent processes contribute to the film growth, as shown in Figure 7:

- Sputtering of Ag from target with argon ion bombardment
- Plasma polymerization of C_2H_4 precursor (with NH_3 interaction)

Admixture of ammonia in gaseous mixture yielded mainly nitrogen containing functionalities such as amines etc. Thus, Ag/a-C:H:N coatings were deposited on textiles in order to obtain antistatic, bio-molecule-adhesive and bacterial resistant nanocomposites.

The bio-molecule incorporation and antibacterial properties of the coated substrates are strongly influenced by nitrogen and Ag content in the matrix (see Figure 6). In this work it was proved that Ag/a-C:H:N coatings having 17% Ag (0.37 Ag/C) and 30.6% N (0.65 N/C) were able to reduce bacterial adhesion by up to 95% until 24 hours as compared to an external PES fabric control.[36] Furthermore, an additional bactericidal effect in the culture media was observed and can be attributed to the release of Ag ions into the media. Klueh et al. described various mechanisms of silver's effect depending on the concentration of free Ag^+ ions:[34]

- Ag^+ forms insoluble compounds with sulfhydryl groups in the cell wall
- Ag^+ blocks the respiratory chain of bacteria
- Ag^+ enters the cell and binds with bacterial DNA.

Thus, the nanocomposites can be used to inhibit the growth of bacteria, fungi, mildew, and odor.[39] The deposited biologically active thin films can be used for multifunctional healthcare interior textiles. These films could be used in tissue engineering such as surface modification of bone implants etc. Furthermore, the metalized-composites produce fabrics with a unique appearance, thus indicating potential candidates for functional and decorative effects for textiles and garments.[40]

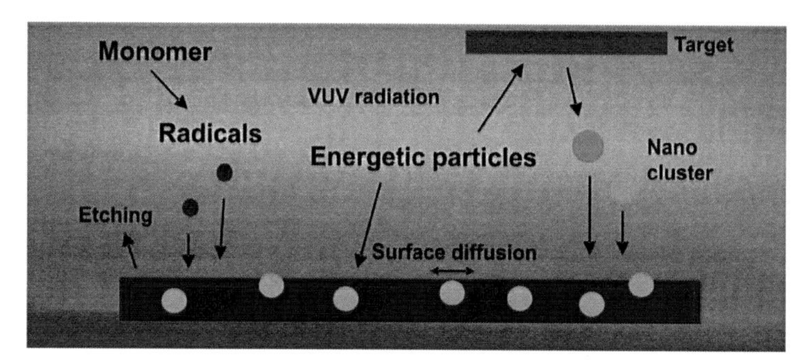

Figure 7. Growth mechanism of two concurrent processes of plasma polymerization and co-sputtering.

10.6. General Conclusion

Plasma processing is a high-technology discipline in materials processing which evolves enormous advantages that are unattainable by strictly wet chemical methods. The plasma surface modification of polymer materials is a useful way to obtain functional polymers by controlling their surface properties without affecting the bulk of polymers.

Surface modification of textiles can be effectively performed, and surface properties strongly depend on the gaseous mixtures used. The surfaces of PET fabrics and foils were modified by low pressure RF plasmas with air, CO_2, water vapor as well as Ar/O_2 and He/O_2 mixtures. In order to increase the wettability of the fabrics, the plasma processing parameters were optimized by means of a suction test with water. It was found that low pressure (10-16 Pa) and medium power (10-16 W) yielded a good penetration of plasma species in the textile structure for all oxygen containing gases and gaseous mixtures used. While the wettability of the PET fabric was increased in all cases, the Ar/O_2 plasma revealed the best hydrophilization effect with respect to water suction and aging. The hydrophilization of PET fabrics was closely related to the surface oxidation. Static and advancing CAs were determined from the capillary rise with water. Both wetting and aging demonstrated a good comparability between plasma-treated PET fabrics and foils, thus indicating a uniform treatment.

Since textiles have heterogeneous complex structures as compared to solid polymeric materials, the modification of a textile surface is not only dependent on plasma process parameters, but also strongly influenced by the textile structure and construction. Moreover, the degree of plasma species penetration and the stability of treatment are closely linked to weave construction. Chapter 4 investigated the influence of plasma activation on textile structures to improve wettability of PES fabrics, and the modification is characterized by CAs (static and advancing) using capillary rise tests with water. The hydrophilic modification is carried out using oxygen containing gaseous mixtures (Ar/O_2 and He/O_2) with long-lasting free radical lifetimes. In all cases the wettability of plasma-treated PES fabrics is improved significantly due to the formation of polar groups on the surface. In particular, the hydrophilicity of looser structured fabrics is improved remarkably as compared to tightly woven fabrics. Furthermore, the capillary phenomenon in fibrous assemblies is also described in this chapter.

In plasma polymerization often additional, non-polymerizable gases are used, either as carrier gas or as reactive co-monomer. The gas ratio represents an additional parameter besides power input, gas flow, pressure, temperature, potentials, and reactor geometry, which complicates the understanding of plasma polymerization processes yielding a controversial discussion of deposition-supporting and etching effects. Therefore, as stated in Chapter 5, the deposition rates of different gas mixtures (O_2/HMDSO, N_2 as well as NH_3/C_xH_y, and inert gas/C_xH_y has thoroughly been investigated on the basis of the following macroscopic approach. Regarding the mass deposition rates depending on the specific energy W/F the radical-dominated plasma polymerization regime for the corresponding polymerizable monomer can be identified by introduction of a modified flow $F = F_m + a\ F_c$ (sum of monomer flow F_m and carrier gas flow F_c with a flow factor a). Any deviations discovered during application of this novel approach indicate additional, e.g. ion-induced, effects. Thus, the understanding of plasma polymerization can be improved using plasma chemistry determined by the activation energy derived from the evaluation of the deposition rates.

It has long been recognized that dyes other than disperse dyes would play a much larger industrial role if they could be applied on PET fabrics at low temperatures. Chapter 6 describes a new process for the dyeing of hydrophobic PET with hydrophilic acid dyestuffs; the process uses an NH_3/C_2H_2 gaseous plasma to modify the fiber surface, which can be considered a pretreatment of PET. The color strength of dyed PET is evidently improved. The dyeability strongly depends on the plasma exposure time, gaseous mixture and energy input W/F. The good fastness properties of dyed PET proved the permanency of the bond between dye molecules and the plasma film. The coatings adhere well to the substrates which were examined by abrasion testing.

Chapter 7 explored the work developing an original process of plasma deposition of nanoporous functionalized coatings, which were applied to PES textiles enabling substrate independent surface dyeing. This novel approach is essentially based on a fine control of rivaling deposition/etching processes during plasma co-polymerization of ammonia with hydrocarbons at low temperature. A nanoporous structure with a large surface area was achieved that contained functional groups inside the coating volume, which were accessible to dye molecules, thus facilitating substrate independent surface dyeing. Depending on the gas ratio, a crosslinked and branched hydrocarbon network yielded excellent mechanical properties and a durable coating as indicated by the high color fastness to washing of the dyed textiles. The dyeing time can be reduced remarkably due to the porous nanoscaled coatings. Since amine-embedded functionalized films were deposited, it was possible for acid dyeing to be applied to the PES at low temperatures.

The plasma-assisted process presented in the Chapter 8 facilitates the deposition of nitrogen containing functional coatings on textiles. The coloration of plasma polymers provides information about the accessible amine functionalities, coating purity and uniformity. RF plasma was used to deposit multi-functional thin films, which show high amines in content. Alternately C_2H_2 and C_2H_4 were mixed with NH_3 to obtain a crosslinked structure that contained functional groups, which were accessible for dye molecules throughout the film volume. Varying deposition conditions were applied to compare hydrocarbon gas mixtures regarding deposition rates, water CAs, aging, dyeability, yellowness index, and rub/wash fastnesses. The deposition rate was found to be higher with C_2H_2 discharges and decreased with increasing ammonia-to-hydrocarbon ratios indicating etching effects, while a permanent hydrophilization could be obtained. Dyeing of the plasma coatings by acid dyestuffs showed that the relative color strength value, i.e. amine functionalities, can be noticeably enhanced, while being strongly influenced by energy input *W/F* and gas ratio. It was evident that the coating quality could be improved significantly using a C_2H_4/NH_3 plasma due to reduced unsaturated bonds, which was investigated by CIELAB color spaces. A high dyeing fastness indicated a strong dye-molecule bonding indicating permanency of the amines.

In order to obtain a permanent hydrophilic modification of material surfaces, the deposition of nanoscaled functional coatings is required with defined process conditions. The incorporation of nitrogen and the generation of free radicals in a-C:H films yields a strong hydrophilic modification, when the deposited surface is exposed to atmosphere. Chemical analysis indicated that the formation of nitrogen functionalities, depending on the NH_3 to hydrocarbon ratio, is mostly due to a replacement of carbon in a-C:H:N films. AFM analysis revealed that nitrogen incorporation plays an important role to obtain nanoporous and crosslinked films. In addition, incorporation of accessible amine functionalities within the coating demonstrated the dyeability of the coatings. This novel combination of polar groups with a suitable texturing realized within crosslinked a-C:H:N coatings proved to be an efficient method providing a long-term mechanical stability of the superhydrophilic coating. This approach represents a step forward compared to known processes such as plasma activation and surfactant-coating methods.

Since nanoporous coatings ensure enlargement of surface area, the coatings can be used, depending on the gas ratios, to improve interfacial adhesion and strength in fiber-reinforced composites. The coatings contain a high number of functional groups which can chemically bind with the matrix materials. Thus, the interfacial surface properties of composites were enhanced using nanoporous coated fibers as compared to untreated fibers.

These findings demonstrate that plasma polymerization provides an eco-friendly, multi-functionalizing surface modification, since the use of chemicals, waste water etc. can be eliminated. Low pressure discharge was performed for the modification which avoids heat generation to the surface and additionally, it delivers high activation energy. The nanoporous plasma polymers can be effectively used as a foundation for multifunctional applications. Thus, the developed nanoporous coatings that incorporate accessible functional groups are the most promising candidates for added-valued textiles explored in this dissertation.

10.7. References

[1] G. Reinert, F. Fuso, R. Hilfiker, E. Schmidt, *Text. Chem. Color.* **1997**, *29 (12)*, 36.
[2] Y. W. H. Wong, C. W. M. Leung, S. K. A. Ku, H. L. I. Lam, *AUTEX Res. J.* **2006**, *6 (1)*, 1.
[3] B. Mahltig, H. Böttcher, K. Rauch, U. Dieckmann, R. Nitsche, T. Fritz, *Thin Solid Films* **2005**, *485*, 108.
[4] J. J. Lee, H. H. Lee, S. I. Eom, J. P. Kim, *Color. Technol.* **2001**, *117*, 134.
[5] F. A. Bottino, G. Di Pasquale, A. Pollicino, A. Recca, *J. Appl. Polym. Sci.* **1998**, *69*, 1251.
[6] E. G. Tsatsaroni, I. C. Eleftheriadis, *Dyes Pigments* **2004**, *61*, 141.
[7] R. Suchentrunk, H. J. Fuesser, G. Staudigl, D. Jonke, M. Meyer, *Surf. Coat. Technol.* **1999**, *112*, 351.
[8] S. Luo, W. J. van Ooij, *J. Adhes. Sci. Technol.* **2002**, *16 (13)*, 1715.
[9] J. Müssig, *Proc. 4th International Wood and Natural Fiber Composite Symposium*, Kassel/Germany, April 10-11, **2002**.
[10] S. Alalto, P. Haapanen, A. Laine, M. Mutanen, J. Toivonen, *Proc. 4th International Wood and Natural Fiber Composite Symposium*, Kassel/Germany, April 10-11, **2002**.
[11] J. Müssig, *Europäsche Akademie*, Newsletter, Akademie-Brief **2006**, *67*, 1.
[12] T. Peijs, *Proc. 4th International Wood and Natural Fiber Composite Symposium*, Kassel/Germany, April 10-11, **2002**.
[13] W. J. van Ooij, S. Luo, S. Datta, *Plasma Polym.* **1999**, *4 (1)*, 33.
[14] Y. J. Hwang, Y. Qiu, C. Zhang, B. Jarrard, R. Stedeford, J. Tsai, Y. C. Park, M. McCord, *J. Adhes. Sci. Technol.* **2003**, *17 (6)*, 847.
[15] J. Müssig, S. Rau, A. S. Herrmann, *J. Natural Fibers* **2006**, *3 (1)*, 59.
[16] R. A. Shanks, A. Hodzic, D. Ridderhof, *J. Appl. Polym. Sci.* **2006**, *99*, 2305.
[17] S. Thomas, *Proc. 4th International Wood and Natural Fiber Composite Symposium*, Kassel/Germany, April 10-11, **2002**.
[18] S. M. Zhang, J. Liu, W. Zhou, L. Cheng, X. D. Guo, *Curr. Appl. Phys.* **2005**, *5*, 516.
[19] D. Plackett, *J. Polym. Environ.* **2004**, *12 (4)*, 131.
[20] R. R. Deshmukh, N. V. Bhat, *Mat. Res. Innovat.* **2003**, *7*, 283.
[21] R. J. Smiley, W. N. Delgass, *J. Mat. Sci.* **1993**, *28*, 3601.
[22] Y. Z. Wan, Y. L. Wang, X. H. Xu, Q. Y. Li, *J. Appl. Polym. Sci.* **2000**, *82*, 150.
[23] A. S. Herrmann, J. Nickel, U. Riedel, *Polym. Degrad. Stabil.* **1998**, *59 (1-3)*, 251.
[24] A. Awal, G. Cescutti, J, Müssig, *J. Appl. Polym. Sci.* **2007**, submitted.
[25] J. M. Park, D. S. Kim, S. R. Kim, *Compos. Sci. Technol.* **2004**, *64*, 847.
[26] M. Creatore, P. Favia, G. Tenuto, A. Valentini, R. d`Agostino, *Plasmas Polym.* **2000**, *5 (3/4)*, 201.
[27] A. Belkind, A. Freilich, J. Lopez, Z. Zhao, W. Zhu, K. Becker, *New J. Phys.* **2005**, *7*, 90.
[28] W. D. Westwood, "*Sputter Deposition*", AVS Education Committee Book Series, Vol. 2, **2003**, New York, pp. 3-5.
[29] B. Chapman, "*Glow Discharge Processes*", John Wiley & Sons, New York, **1980**, pp. 177-184.
[30] J. Müssig, G. Cescutti, D. Hegemann, M. M. Hossain, T. Donath, F. Beckmann, *HASYLAB Annual Report 2006.* Hamburg: Hamburger

Synchrotonstrahungslabor HASYLAB am Deutschen Elektronen-Synchroton DESY in der Helmholtz-Gemeinschaft HGF, **2006**, pp. 631-632.
[31] G. Cescutti, J. Müssig, K. Specht, A. K. Bledzki, *Proc. 6th Global Wood and Natural Fiber Symposium*, Kassel/Germany, **2006**, pp. B9-1 to B9-11.
[32] J. Müssig, H. G. Schmid, *Microscopy and Microanalysis* **2004**, *10 (2)*, pp. 1332CD-1333CD.
[33] T. Walther, T. Donath, K. Terzic, H. Meine, H. Thömen, F. Beckmann, *HASYLAB Annual Report* **2006**, pp. 455-456.
[34] U. Klueh, V. Wagner, S. Kelly, A. Johnson, J. D. Bryers, *J. Biomed. Mater. Res. A* **2000**, *53*, 621.
[35] V. N. Golubovich, I. L. Rabotnova, *Microbiololgy+* **1974**, *43*, 948.
[36] D. J. Balazs, M. M. Hossain, E. Brombacher, G. Fortunato, E. Körner, D. Hegemann, *Plasma Process Polym.* **2007**, *4*, S380.
[37] S. Ido, M. Kashiwagi, M. Takahashi, *Jpn. J. Appl. Phys.* **1999**, *38*, 4450.
[38] H. Koshelyev, *Proc. 4th International Workshop and School*, Kudowa Zdroj/Poland, June 7-13, **2004**.
[39] M. Gurian, *J. Coated Fabrics* **1995**, *25*, 13.
[40] S. Q. Jiang, E. Newton, C. W. M. Yuen, C. W. Kan, *J. Appl. Polym. Sci.* **2005**, *96*, 919.

CURRICULUM VITAE

Mohammad Mokbul Hossain
Born on December 15, 1976 in Dhaka, Bangladesh

Professional Experiences

01/2008-present	**Manager**, Dept. of Plasma Technology, Textilveredlung Grabher GmbH, Lustenau, Austria.
09/2007-12/2007	**Postdoc,** Plasma-modified Surfaces, Empa, Swiss Materials Science & Technology, St. Gallen, Switzerland.
03/2004-08/2007	**Doctoral student,** Plasma-modified Surfaces, Empa, Swiss Materials Science & Technology, St. Gallen, Switzerland.
05/2003-09/2003	**Master's student,** Dresden University of Technology, Dresden, Germany.
09/1999-03/2001	**Laboratory In-charge,** Capital Mercury Apparel Ltd., Dhaka, Bangladesh
01/1999-09/1999	**Laboratory Technologist**, Opex Garments Ltd., Dhaka, Bangladesh

Industrial Interns

03/2003-04/2003	Development of a recipe for the pretreatment of ink jet textile printing, BASF AG, Ludwigshafen, Germany
July 15-22, 2000	Test of textile fabrics, Beximco Textiles Ltd. Dhaka, Bangladesh
10/1998-11/1998	Mid course industrial training, Mohammadi Group, Bangladesh

Education

03/2004-08/2007	**Ph.D. thesis: Plasma technology for deposition and surface modification** Empa, Swiss Materials Science & Technology (ETH-domain), St. Gallen, Switzerland; University of Bremen, Bremen, Germany ("magna cum laude")
2001-2003	**M.Sc. in Textile Engineering (M.Sc. Eng.)**, Faculty of Mechanical Engineering, Dresden University of Technology, Germany (Thesis: "Note 1" & overall grade: "Note 2")
1993-1997	**B.Sc. in Textile Technology**, University of Dhaka, Bangladesh (1st Division)
1991-1993	**H.S.C.**, Ananda Mohan College, Bangladesh (1st Division)
1981-1991	**S.S.C.**, Bhaighat High School, Bangladesh (1st Division)

Scholarships and Awards

DAAD scholarship, Granted a full scholarship for postgraduate course on "M.Sc. in Textile Engineering" by DAAD. (04/2001-09/2003)

Merit scholarship, Awarded for a four-academic-year scholarship at University of Dhaka, Bangladesh. (10/1993-09/1997)

BGMEA award, Scholarship award for the course in "Business Communication Skills" at the British Council, Dhaka which was organized by Bangladesh Garment Manufacturer & Exporter Association (BGMEA). (11/1998-01/1999)

Special Courses and Trainings

"**Basics in Management**" at Empa Duebendorf which was organized by Empa, Swiss Materials Science & Technology, July 17-21, 2006.

"**Special Topics in Using Scientific Databases**" at Empa St. Gallen which was organized by Empa Section Technology Transfer and Knowledge Management, August 31, 2005.

"**How to Write a Good Scientific Paper**" at Empa Duebendorf which was organized by Empa, Swiss Materials Science & Technology, August 24, 2004.

"**ImageAccess**" at Empa St. Gallen which was organized by Imagic Bildverarbeitung AG, Glattbrugg, March 15-16, 2004.

"**Writing English for Science**" at EAWAG (Swiss Federal Institute for Environmental Science and Technology) Duebendorf which was organized by EAWAG, March 1, 3, 7, 10, 2004.

"**Color Evaluation Program**" at Kowloon, Hong Kong which was organized by Capital Mercury Apparel Ltd. & Data Color International (USA), July 27, 2000.

Publications Related to Thesis

[1] **M. M. Hossain**, J. Müssig, A. S. Herrmann, D. Hegemann, "Ammonia/Acetylene Plasma Deposition: An Alternative Approach to Dyeing of Poly(ethylene terephthalate) (PET) Fabrics at Low Temperature" *J. Appl. Polym. Sci.* 2009, *111*, 2545.

[2] **M. M. Hossain**, D. Hegemann, M. Amberg, J. Müssig, "Plasma Polymerization for Deposition of Nanoporous Multifunctional Thin Films on Textiles", *Proc. AUTEX Technical 2007 Conference*, June 26-28, 2007, Tampere, Finland.

[3] **M. M. Hossain**, D. Hegemann, A. S. Herrmann, "Plasma Deposition of Amine-embedded Nanoporous Ultrathin Films on Polyester – Enables Substrate Independent Surface Dyeing" *Plasma Process. Polym.* 2007, *4 (S1)*, S1068.

[4] **M. M. Hossain**, D. Hegemann, A. S. Herrmann, G. Fortunato, M. Heuberger, "Plasma Deposition of Permanent Superhydrophilic a-C:H:N Films on Textiles" *Plasma Process. Polym.* 2007, *4*, 471.

[5] D. J. Balazs, **M. M. Hossain**, E. Brombacher, G. Fortunato, E. Körner, D. Hegemann, "Multifunctional Nano-composite Plasma Coatings – Enabling New Applications in Biomaterials" *Plasma Process. Polym.* 2007, *4 (S1)*, S380.

[6] **M. M. Hossain**, A. S. Herrmann, D. Hegemann, "Incorporation of Accessible Functionalities in Nanoscaled Coatings on Textiles Characterized by Coloration" *Plasma Process. Polym.* 2007, *4*, 135.

[7] D. Hegemann, **M. M. Hossain**, D. J. Balazs, "Nanostructured Plasma Coatings to Obtain Multifunctional Textile Surfaces" *Prog. Org. Coat.* 2007, *58*, 237.

[8] D. Hegemann, **M. M. Hossain**, E. Körner, D. J. Balazs, "Macroscopic Description of Plasma Polymerization" *Plasma Process. Polym.* 2007, *4 (3)*, 229.

[9] **M. M. Hossain**, D. Hegemann, A. S. Herrmann, P. Chabrecek, “Contact Angle Determination of Plasma-Treated Poly(ethylene terephthalate) Fabrics and Foils” *J. Appl. Polym. Sci.* 2006, *102*, 1452.

[10] **M. M. Hossain**, D. Hegemann, A. S. Herrmann, “Plasma Hydrophilization Effect on Different Textile Structures” *Plasma Process. Polym.* 2006, *3*, 299.
This article was one of the **most downloaded** articles published between November 2005 and October 2006, *Plasma Process. Polym.* 2007, *4*, 11.

[11] D. Hegemann, **M. M. Hossain**, “Influence of Non-polymerizable Gases Added During Plasma Polymerization” *Plasma Process. Polym.* 2005, *2*, 554.

Oral Conference and Workshop Contributions

M. M. Hossain, D. J. Balazs, S. Lisher, K. Grieder, G. Fortunato, E. Koerner, P. Wick, M. Heuberger “Multifunctional Plasma Coatings – Enabling New Biomaterials Applications”, MRS (Materials Research Society) Fall Meeting", November 26-30, 2007, Boston, USA.

M. M. Hossain, D. Hegemann, “Multifunctional Nanoporous Plasma Coatings”, NanoEurope, September 11-13, 2007, St. Gallen, Switzerland.

M. M. Hossain, D. Hegemann, J. Müssig, A. S. Herrmann “Nanoporous Coatings Deposited by RF Plasma - Improving Interfacial Properties in Fiber-reinforced Composites”, 28th Risø International Symposium on Materials Science: "Interface Design of Polymer Matrix Composites - Mechanics, Chemistry, Modelling and Manufacturing", September 3-6, 2007, Roskilde, Denmark.

M. M. Hossain, D. Hegemann, M. Amberg, J. Müssig, “Plasma Polymerization for Deposition of Nanoporous Multifunctional Thin Films on Textiles”, AUTEX Technical 2007 Conference, June 26-28, 2007, Tampere, Finland.

M. M. Hossain, D. Hegemann, M. Heuberger, “Wash Permanent Superhydrophobic Coatings on Textiles by RF Plasma” Textile Workshop, Empa, Swiss Materials Science & Technology, January 16, 2007, St. Gallen, Switzerland.

M. M. Hossain, D. Hegemann “Plasma Deposition of Nanoporous Coatings for Multifunctional Textiles”, Fiber Society Technical Conference, October 10-13, 2006, Knoxville, USA.

M. M. Hossain, D. Hegemann, A. S. Herrmann, “Nanoporous Functionalized Ultrathin Films” Textile Workshop, Empa, Swiss Materials Science & Technology, June 12, 2006, St. Gallen, Switzerland.

M. M. Hossain, D. Hegemann, A. S. Herrmann, “Hydrophilic Modification of Textiles” Textile Workshop, Empa, Swiss Materials Science & Technology, August 24, 2005, St. Gallen, Switzerland.

M. M. Hossain, D. Hegemann, A. S. Herrmann, “Functionalization of Textiles by Plasma Technology”, Textile Workshop, Empa, Swiss Materials Science & Technology, October 25, 2004, St. Gallen, Switzerland.

M. M. Hossain, D. Hegemann, A. S. Herrmann, “Surface Modification of Materials by Plasma Technology”, Fiber Institute Bremen, University of Bremen, April 11, 2004, Bremen, Germany.

Poster Presentations

S. Guimond, **M. M. Hossain**, and D. Hegemann, "Nanostructured Plasma-Polymer Coatings for Textile Applications”, New Perspectives of Plasma Science and Technology, October 23 – 25, 2007, Brno, Czech Republic.

M. M. Hossain, D. Hegemann, G. Günter, M. Heuberger, “Nanoporous Plasma Coatings – a Platform for Multifunctional Applications”, Empa PhD Symposium, November 21, 2007, Duebendorf, Switzerland.
D. Hegemann, **M. M. Hossain**, E. Körner, and D. J. Balazs "Nanostructured Plasma Coatings for Multifunctional Textiles", NanoEurope Conference 2007, September 11-13, 2007, St. Gallen, Switzerland.
M. M. Hossain, D. Hegemann, A. S. Herrmann, “Plasma Deposition of Amine-embedded Nanoporous Ultrathin Films on Polyester – Enables Substrate Independent Surface Dyeing”, Plasma Surface Engineering Conference, September 10-15, 2006, Garmisch-Partenkirchen, Germany.
D. Hegemann, **M. M. Hossain**, E. Körner, D. J. Balazs "Nanostructured Plasma Coatings for Multifunctional Textiles", NanoEurope Fair & Conference, September 11-13, 2006, St. Gallen, Switzerland.
D. Hegemann, **M. M. Hossain**, E. Bertaux, A. Ritter, “Combining the Impossible: Excellent Mechanical Stability of Permanent Non-wetting Coatings” by NanoEurope Fair & Conference, September 11-13, 2006, St. Gallen, Switzerland.
M. M. Hossain, D. Hegemann, A. S. Herrmann, “Contact Angle Determination on Plasma-Treated PET Fabrics and Foils”, Fiber Society Spring Conference, May 25-27, 2005, St. Gallen, Switzerland.